AF513402

Cell Division in Plants

Cell Division in Plants

Nirmay Adiga

RANDOM

RANDOM PUBLICATIONS
NEW DELHI (INDIA)

Cell Division in Plants

ISBN 978-93-5111-371-3
© Reserved

All Rights Reserved. No Part of this book may be reproduced in any manner without written permission.

Published in 2014 in India by

RANDOM PUBLICATIONS

Reprint, 2017

4376-A/4B, Gali Murari Lal, Ansari Road

New Delhi-110 002

Phone : +91-11-43580356, +91-11-23289044

e-mail: randomexports@gmail.com, sales@randompublications.com, info@randompublications.com

Type Setting by: Friends Media, Delhi-110089

Digitally Printed at: Replika Press Pvt. Ltd.

Preface

Plant cells are eukaryotic cells that differ in several key aspects from the cells of other eukaryotic organisms. During plant development, cell walls ensure that the relative positions of cells change little, if any. Consequently, the cellular organization of a mature plant tissue closely reflects the pattern of cell division during its development. Cell division is the process by which a parent cell divides into two or more daughter cells. A cell division is usually a small segment of a larger cell cycle. This type of division in eukaryotes is known as mitosis. The division of eukaryotic cells involves the assembly of complex cytoskeletal structures to exert the forces required for chromosome segregation and cytokinesis. In plants, empirical evidence suggests that tensional forces within the cytoskeleton cause cells to divide along the plane that minimizes the surface area of the cell plate (Errera's rule) while creating daughter cells of equal size. However, exceptions to Errera's rule cast doubt on whether a broadly applicable rule can be formulated for plant cell division.

Cell division is characterised by the segregation of genetic material and the redistribution of cellular contents facilitated by the cytoskeleton. Plant cells derive from divisions in the meristems of developing organs. Moreover, a comparison across land plants and their most recent algal ancestors confirms that the probability distribution is widely conserved and independent of cell shape and size. Cell division does not have major differences between what happens in animals and what happens in plants, although there are still important differences that must be acknowledged. Cells in both animals and plants undergo a process called mitosis. But in that process, animals reproduce cells all the time throughout their bodies. Plants reproduce by cell division in specific regions. Mitosis is necessary for the maintenance of the chromosome sets in all species of animals and plants.

The purpose of the book is to emphasise the structural and spatial events which occur during the development of specialised plant cells. It is designed to fill the gap between descriptive anatomy books on the one hand and purely physiological books on the other. Its novelty is in its emphasis on the interaction between the structure of a plant cell and the way in which it performs its role in the plant.

I would like to thank my team for standing beside me throughout my career and writing this book. My special thanks go to "Random Publications" who have published the book.

—Nirmay Adiga

Contents

1

Introduction

Cell biology is a scientific discipline that studies cells – their physiological properties, their structure, the organelles they contain, interactions with their environment, their life cycle, division and death. This is done both on a microscopic and molecular level. Cell biology research encompasses both the great diversity of single-celled organisms like bacteria and protozoa, as well as the many specialized cells in multicellular organisms such as humans, plants, and sponges.

Knowing the components of cells and how cells work is fundamental to all biological sciences. Appreciating the similarities and differences between cell types is particularly important to the fields of cell and molecular biology as well as to biomedical fields such as cancer research and developmental biology. These fundamental similarities and differences provide a unifying theme, sometimes allowing the principles learned from studying one cell type to be extrapolated and generalized to other cell types. Therefore, research in cell biology is closely related to genetics, biochemistry, molecular biology, immunology, and developmental biology.

Processes

Movement of Proteins

Each type of protein is usually sent to a particular part of the cell. An important part of cell biology is the investigation of molecular mechanisms by which proteins are moved to different places inside cells or secreted from cells. Most proteins are synthesized by ribosomes in the rough endoplasmic reticulum (rER). Ribosomes contain the nucleic acid RNA, which assembles and joins amino acids to make proteins. They can be found alone or in groups within the cytoplasm as well as on the rER. This process is known as protein biosynthesis. Biosynthesis (also called biogenesis) is an enzyme-catalyzed process in cells of living organisms by which substrates are converted to more complex products (also simply known as protein translation). Some proteins, such as those to be incorporated in membranes (known as membrane proteins), are transported into the rough endoplasmic reticulum (ER) during synthesis. This process can be followed by transportation and processing in the Golgi apparatus.

The Golgi apparatus is a large organelle that processes proteins and prepares them for use both inside and outside the cell. The Golgi apparatus is somewhat like a post office. It receives items (proteins from the ER), packages and labels them, and then sends them on to their destinations (to different parts of the cell or to the cell membrane for transport out of the cell). From the Golgi, membrane proteins can move to the plasma membrane, to other sub-cellular compartments, or they can be secreted from the cell. The ER and Golgi can be thought of as the "membrane protein synthesis compartment" and the "membrane protein processing compartment", respectively. There is a semi-constant flux of proteins through these compartments. ER and Golgi-resident proteins associate with other proteins but remain in their respective compartments. Other proteins "flow" through the ER and Golgi to the plasma membrane. Motor proteins transport membrane protein-containing vesicles along cytoskeletal tracks to distant parts of cells such as axon terminals.

Some proteins that are made in the cytoplasm contain structural features that target them for transport into mitochondria or the nucleus. Some mitochondrial proteins are made inside mitochondria and are coded for by mitochondrial DNA. In plants, chloroplasts also make some cell proteins.

Extracellular and cell surface proteins destined to be degraded can move back into intracellular compartments upon being incorporated into endocytosed vesicles some of which fuse with lysosomes where the proteins are broken down to their individual amino acids. The degradation of some membrane proteins begins while still at the cell surface when they are separated by secretases. Proteins that function in the cytoplasm are often degraded by proteasomes.

Other Cellular Processes

- Active transport and Passive transport - Movement of molecules into and out of cells.

- Autophagy - The process whereby cells "eat" their own internal components or microbial invaders.
- Adhesion - Holding together cells and tissues.
- Reproduction - Made possible by the combination of sperm made in the testiculi (contained in some male cells' nuclei) and the egg made in the ovary (contained in the nucleus of a female cell). When the sperm breaks through the hard outer shell of the egg a new cell embryo is formed, which, in humans, grows to full size in 9 months.
- Cell movement: Chemotaxis, Contraction, cilia and flagella.
- Cell signalling - Regulation of cell behaviour by signals from outside.
- DNA repair, Cell death and Cell senescence
- Metabolism: Glycolysis, respiration, Photosynthesis
- Transcription and mRNA splicing - gene expression.

Internal Cellular Structures

- Chloroplast - key organelle for photosynthesis (only found in plant cells)
- Cell wall - extra layer of protection (only found in plant cells)
- Cell membrane - the part of the cell which separates the cells from the outside environment and protects the cell, as well as regulating what goes in and out of the cell
- Cilium - motile microtubule-containing structure of eukaryotes
- Cytoplasm - contents of the main fluid-filled space inside cells. Any chemical reactions also happen in here.
- Cytoskeleton - protein filaments inside cells
- Endoplasmic reticulum - major site of membrane protein synthesis
- Flagellum - motile structure of bacteria, archaea and eukaryotes
- Golgi apparatus - site of protein glycosylation in the endomembrane system
- Lipid bilayer - fundamental organizational structure of cell membranes
- Lysosome - break down cellular waste products and debris into simple compounds (only found in animal cells)
- Membrane lipid and protein barrier

- Mitochondrion - major energy-producing organelle by releasing it in the form of ATP
- Nucleus - holds most of the DNA of eukaryotic cells and controls all cellular activities
- Organelle - term used for major subcellular structures
- Ribosome - RNA and protein complex required for protein synthesis in cells
- Vesicle - small membrane-bounded spheres inside cells

Techniques Used to Study Cells

Cells may be observed under the microscope. This includes the Optical Microscope, Transmission Electron Microscope, Scanning Electron Microscope, Fluorescence Microscope, and by Confocal Microscopy.

Several different techniques exist to study cells are

- Cell culture is the basic technique of growing cells in a laboratory independent of an organism.
- Immunostaining, also known as immunohistochemistry, is a specialized histological method used to localize proteins in cells or tissue slices. Unlike regular histology, which uses stains to identify cells, cellular components or protein classes, immunostaining requires the reaction of an antibody directed against the protein of interest within the tissue or cell. Through the use of proper controls and published protocols (need to add reference links here), specificity of the antibody-antigen reaction can be achieved. Once this complex is formed, it is identified via either a "tag" attached directly to the antibody, or added in an additional technical step. Commonly used "tags" include fluorophores or enzymes. In the case of the former, detection of the location of the "immuno-stained" protein occurs via fluorescence microscopy. With an enzymatic tag, such as horse radish peroxidase, a chemical reaction is carried out that results in a dark colour in the location of the protein of interest. This darkened pattern is then detected using light microscopy.
- Computational genomics is used to find patterns in genomic information
- DNA microarrays identify changes in transcript levels between different experimental conditions.
- Gene knockdown mutates a selected gene.

- In situ hybridization shows which cells are expressing a particular RNA transcript.
- PCR can be used to determine how many copies of a gene are present in a cell.
- Transfection introduces a new gene into a cell, usually an expression construct

Purification of cells and their parts Purification may be performed using the following methods:

- Cell fractionation
  - o Release of cellular organelles by disruption of cells.
  - o Separation of different organelles by centrifugation.
- Flow cytometry
- Immunoprecipitation
- Proteins extracted from cell membranes by detergents and salts or other kinds of chemicals.

Plant Cell

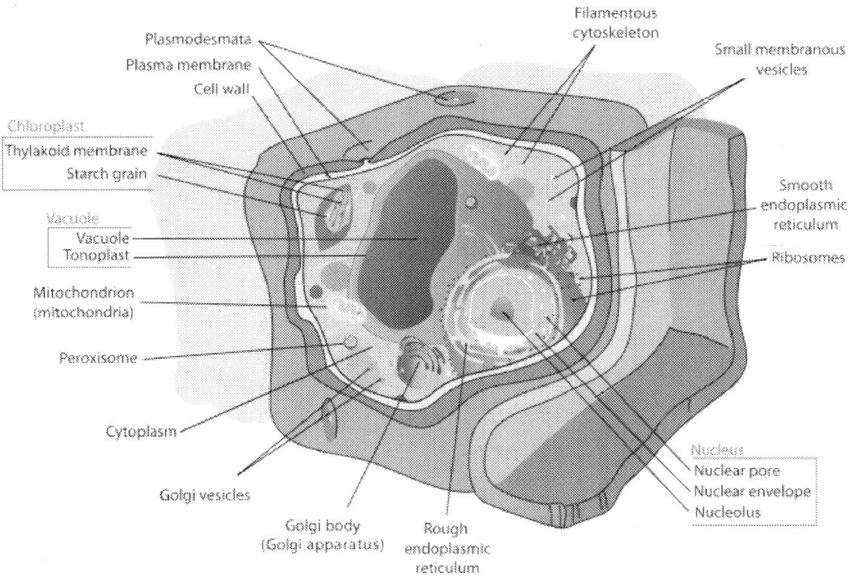

Figure: *Plant cell structure*

Plant cells are eukaryotic cells that differ in several key aspects from the cells of other eukaryotic organisms. Their distinctive features include:

- A large central vacuole, a water-filled volume enclosed by a membrane known as the *tonoplast* that maintains the cell's turgor, controls movement of molecules between the cytosol and sap, stores useful material and digests waste proteins and organelles.

- A cell wall composed of cellulose and hemicellulose, pectin and in many cases lignin, is secreted by the protoplast on the outside of the cell membrane. This contrasts with the cell walls of fungi (which are made of chitin), and of bacteria, which are made of peptidoglycan.
- Specialized cell-to-cell communication pathways known as plasmodesmata, pores in the primary cell wall through which the plasmalemma and endoplasmic reticulum of adjacent cells are continuous.
- Plastids, the most notable being the chloroplast, which contains chlorophyll, a green-coloured pigment that absorbs sunlight, and allows the plant to make its own food in the process known as photosynthesis. Other types of plastids are the amyloplasts, specialized for starch storage, elaioplasts specialized for fat storage, and chromoplasts specialized for synthesis and storage of pigments. As in mitochondria, which have a genome encoding 37 genes, plastids have their own genomes of about 100–120 unique genes and, it is presumed, arose as prokaryotic endosymbionts living in the cells of an early eukaryotic ancestor of the land plants and algae.
- Cell division by construction of a phragmoplast as a template for building a cell plate late in cytokinesis is characteristic of land plants and a few groups of algae, notably the Charophytes and the Order Trentepohliales
- The sperm of bryophytes and pteridophytes, Cycads and Ginkgo have flagella similar to those in animals, but higher plants, (including Gymnosperms and flowering plants) lack the flagellae and centrioles that are present in animal cells.

Cell Types

- Parenchyma cells are living cells that have functions ranging from storage and support to photosynthesis and phloem loading (transfer cells). Apart from the xylem and phloem in their vascular bundles, leaves are composed mainly of parenchyma cells. Some parenchyma cells, as in the epidermis, are specialized for light penetration and focusing or regulation of gas exchange, but others are among the least specialized cells in plant tissue, and may remain totipotent, capable of dividing to produce new populations of undifferentiated cells, throughout their lives. Parenchyma cells have thin, permeable primary walls enabling the transport of small molecules between them, and their cytoplasm is responsible for a wide range of biochemical functions such as nectar secretion, or the manufacture of secondary products that discourage herbivory. Parenchyma cells that contain many chloroplasts and are concerned primarily with photosynthesis are called chlorenchyma cells. Others, such as the majority of the parenchyma cells in potato tubers and the seed cotyledons of legumes, have a storage function.

- Collenchyma cells – collenchyma cells are alive at maturity and have only a primary wall. These cells mature from meristem derivatives that initially resemble parenchyma, but differences quickly become apparent. Plastids do not develop, and the secretory apparatus (ER and Golgi) proliferates to secrete additional primary wall. The wall is most commonly thickest at the corners, where three or more cells come in contact, and thinnest where only two cells come in contact, though other arrangements of the wall thickening are possible.

Pectin and hemicellulose are the dominant constituents of collenchyma cell walls of dicotyledon angiosperms, which may contain as little as 20% of cellulose in *Petasites*. Collenchyma cells are typically quite elongated, and may divide transversely to give a septate appearance. The role of this cell type is to support the plant in axes still growing in length, and to confer flexibility and tensile strength on tissues. The primary wall lacks lignin that would make it tough and rigid, so this cell type provides what could be called plastic support – support that can hold a young stem or petiole into the air, but in cells that can be stretched as the cells around them elongate. Stretchable support (without elastic snap-back) is a good way to describe what collenchyma does. Parts of the strings in celery are collenchyma.

- Sclerenchyma cells – Sclerenchyma cells (from the Greek skleros, *hard*) are hard and tough cells with a function in mechanical support. They are of two broad types – sclereids or stone cells and fibres. The cells develop an extensive secondary cell wall that is laid down on the inside of the primary cell wall. The secondary wall is impregnated with lignin, making it hard and impermeable to water. Thus, these cells cannot survive for long' as they cannot exchange sufficient material to maintain active metabolism. Sclerenchyma cells are typically dead at functional maturity, and the cytoplasm is missing, leaving an empty central cavity.

Functions for sclereid cells (hard cells that give leaves or fruits a gritty texture) include discouraging herbivory, by damaging digestive passages in small insect larval stages, and physical protection (a solid tissue of hard sclereid cells form the pit wall in a peach and many other fruits). Functions of fibres include provision of load-bearing support and tensile strength to the leaves and stems of herbaceous plants. Sclerenchyma fibres are not involved in conduction, either of water and nutrients (as in the xylem) or of carbon compounds (as in the phloem), but it is likely that they may have evolved as modifications of xylem and phloem initials in early land plants.

Tissue Types

The major classes of cells differentiate from undifferentiated meristematic cells (analogous to the stem cells of animals) to form the tissue structures of roots, stems, leaves, flowers, and reproductive structures.

Xylem cells are elongated cells with lignified secondary thickening of the cell walls. Xylem cells are specialised for conduction of water, and first appeared in plants during their transition to land in the Silurian period more than 425 million years ago. The possession of xylem defines the vascular plants or Tracheophytes. Xylem tracheids are pointed, elongated xylem cells, the simplest of which have continuous primary cell walls and lignified secondary wall thickenings in the form of rings, hoops, or reticulate networks. More complex tracheids with valve-like perforations called bordered pits characterise the gymnosperms. The ferns and other pteridophytes and the gymnosperms have only xylem tracheids, while the angiosperms also have xylem vessels. Vessel members are hollow xylem cells without end walls that are aligned end-to-end so as to form long continuous tubes. The bryophytes lack true xylem cells, but their sporophytes have a water-conducting tissue known as the hydrome that is composed of elongated cells of simpler construction.

Phloem is a specialised tissue for food transport in higher plants. Phloem cells mainly transport sucrose along pressure gradients generated by osmosis. This phenomenon is called translocation. Phloem consists of two cell types, the sieve tubes and the intimately associated companion cells. The sieve tube elements lack nuclei and ribosomes, and their metabolism and functions are regulated by the adjacent nucleate companion cells. Sieve tubes are joined end-to-end with perforate end-plates between known as *sieve plates*, which allow transport of photosynthate between the sieve elements. The companion cells, connected to the sieve tubes via plasmodesmata, are responsible for loading the phloem with sugars. The bryophytes lack phloem, but moss sporophytes have a simpler tissue with analogous function known as the leptome.

Plant epidermal cells are specialised parenchyma cells covering the external surfaces of leaves, stems and roots. The epidermal cells of aerial organs arise from the superficial layer of cells known as the *tunica* (L1 and L2 layers) that covers the plant shoot apex, whereas the cortex and vascular tissues arise from innermost layer of the shoot apex known as the *corpus* (L3 layer). The epidermis of roots originates from the layer of cells immediately beneath the root cap.

The epidermis of all aerial organs, but not roots, is covered with a cuticle made of the polyester cutin and/or the hydrocarbon polymer cutan with a superficial layer of waxes. The epidermal cells of the primary shoot are thought to be the only plant cells with the biochemical capacity to synthesize cutin. Several cell types may be present in the epidermis. Notable among these are the stomatal guard cells, glandular and clothing hairs or trichomes, and the root hairs of primary roots. In the shoot epidermis of most plants, only the guard cells have chloroplasts.

Ground Tissue

The ground tissue of plants can be divided into three classes based on the nature of the cell walls. Parenchyma cells have thin primary walls and usually remain alive after they become mature. Parenchyma forms the "filler" tissue in the soft parts of plants. Collenchyma cells have thin primary walls with some areas of secondary thickening. Collenchyma provides extra structural support, particularly in regions of new growth. Sclerenchyma cells have thick lignified secondary walls and often die when mature. Sclerenchyma provides the main structural support to a plant.

Parenchyma

Parenchyma is a versatile ground tissue that generally constitutes the "filler" tissue in soft parts of plants. It forms, among other things, the cortex and pith of stems, the cortex of roots, the mesophyll of leaves, the pulp of fruits, and the endosperm of seeds. Parenchyma cells are living cells and may remain meristematic at maturity—meaning that they are capable of cell division if stimulated. They have thin but flexible cellulose cell walls, and are generally polyhedral when close-packed, but can be roughly spherical when isolated from their neighbours. They have large central vacuoles, which allow the cells to store and regulate ions, waste products, and water. Tissue specialised for food storage is commonly formed of parenchyma cells.

Parenchyma cells have a variety of functions:

- In leaves, they form the mesophyll and are responsible for photosynthesis and the exchange of gases, parenchyma cells in the mesophyll of leaves are a specialised parenchyma cells called chlorenchyma cells (parenchyma cells with chloroplasts).
- Storage of starch, protein, fats, oils and water in roots, tubers (e.g. potatoes), seed endosperm (e.g. cereals) and cotyledons (e.g. pulses and peanuts)
- Secretion (e.g. the parenchyma cells lining the inside of resin ducts)
- Wound repair and the potential for renewed meristematic activity
- Other specialised functions such as aeration (aerenchyma) and support

The shape of parenchyma cells varies with their function. In the spongy mesophyll of a leaf, parenchyma cells range from near-spherical and loosely arranged with large intercellular spaces, to branched or stellate, mutually interconnected with their neighbours at the ends of their arms to form a three-dimensional network, like in the red kidney bean *Phaseolus vulgaris* and other mesophytes. These cells, along with the epidermal guard cells of the stoma, form a system of air spaces and chambers that regulate the exchange of gases. In some works the cells of the leaf epidermis are regarded as specialised parenchymal cells, but the modern preference has long been to classify the epidermis as plant dermal tissue, and parenchyma as ground tissue.

Collenchyma

The first use of "collenchyma" was by Link (1837) who used it to describe the sticky substance on *Bletia* (Orchidaceae) pollen. Complaining about Link's excessive nomenclature, Schleiden (1839) stated mockingly that the term "collenchyma" could have more easily been used to describe elongated sub-epidermal cells with unevenly thickened cells.

Collenchyma tissue is composed of elongated cells with irregularly thickened walls. They provide structural support, particularly in growing shoots and leaves. Collenchyma tissue makes up things such as the resilient strands in stalks of celery. Collenchyma cells are usually living, and have only a thick primary cell wall made up of cellulose and pectin. Cell wall thickness is strongly affected by mechanical stress upon the plant. The walls of collenchyma in shaken plants (to mimic the effects of wind etc.), may be 40–100% thicker than those not shaken.

There are four main types of collenchyma:

- Angular collenchyma (thickened at intercellular contact points)
- Tangential collenchyma (cells arranged into ordered rows and thickened at the tangential face of the cell wall)
- Annular collenchyma (uniformly thickened cell walls)
- Lacunar collenchyma (collenchyma with intercellular spaces)

Collenchyma cells are most often found adjacent to outer growing tissues such as the vascular cambium and are known for increasing structural support and integrity.

Sclerenchyma

Sclerenchyma is the supporting tissue in plants. Two types of sclerenchyma cells exist: fibres and sclereids. Their walls consist of cellulose, hemicellulose and lignin. Sclerenchyma cells are the principal, supporting cells in plant tissues that have ceased elongation. Sclerenchyma fibres are of great economical importance, since they constitute the source material for many fabrics (flax, hemp, jute, ramie).

Unlike the collenchyma, mature sclerenchyma is composed of dead cells with extremely thick cell walls (secondary walls) that make up to 90% of the whole cell volume. The term "sclerenchyma" is derived from the Greek "sklç-rós", meaning "hard". It is the hard, thick walls that make sclerenchyma cells important strengthening and supporting elements in plant parts that have ceased elongation. The difference between fibres and sclereids is not always clear. Transitions do exist, sometimes even within one and the same plant.

Fibres

Fibres or bast are generally long, slender, so-called prosenchymatous cells, usually occurring in strands or bundles. Such bundles or the totality of a stem's bundles are colloquially called fibres. Their high load-bearing capacity and the ease with which they can be processed has since antiquity made them the source material for a number of things, like ropes, fabrics or mattresses. The fibres of flax (*Linum usitatissimum*) have been known in Europe and Egypt for more than 3000 years, those of hemp (*Cannabis sativa*) in China for just as long. These fibres, and those of jute (*Corchorus capsularis*) and ramie (*Boehmeria nivea*, a nettle), are extremely soft and elastic and are especially well suited for the processing to textiles. Their principal cell wall material is cellulose.

Contrasting are hard fibres that are mostly found in monocots. Typical examples are the fibres of many grasses, agaves (sisal: *Agave sisalana*), lilies (*Yucca* or *Phormium tenax*), *Musa textilis* and others. Their cell walls contain, besides cellulose, a high proportion of lignin. The load-bearing capacity of *Phormium tenax* is as high as 20–25 kg/ mm^2, the same as that of good steel wire (25 kg/ mm^2), but the fibre tears as soon as too great a strain is placed upon it, while the wire distorts and does not tear before a strain of 80 kg/mm^2. The thickening of a cell wall has been studied in *Linum*. Starting at the centre of the fibre are the thickening layers of the secondary wall deposited one after the other. Growth at both tips of the cell leads to simultaneous elongation. During development the layers of secondary material seem like tubes, of which the outer one is always longer and older than the next. After completion of growth the missing parts are supplemented, so that the wall is evenly thickened up to the tips of the fibres.

Fibres usually originate from meristematic tissues. Cambium and procambium are their main centres of production. They are usually associated with the xylem and phloem of the vascular bundles. The fibres of the xylem are always lignified, while those of the phloem are cellulosic. Reliable evidence for the fibre cells' evolutionary origin from tracheids exists. During evolution the strength of the tracheid cell walls was enhanced, the ability to conduct water was lost and the size of the pits reduced. Fibres that do not belong to the xylem are bast (outside the ring of cambium) and such fibres that are arranged in characteristic patterns at different sites of the shoot.

Sclereids

Sclereids are a reduced form of sclerenchyma cells with highly thickened, lignified walls. These have a shape of a star. They are small bundles of sclerenchyma tissue in plants that form durable layers, such as the cores of apples and the gritty texture of pears. Sclereids are variable in shape. The cells can be isodiametric, prosenchymatic, forked or elaborately branched. They can be grouped into bundles, can form complete tubes located at the periphery or can occur as single cells or small groups of cells within parenchyma tissues. But compared with most fibres, sclereids are relatively short. Characteristic examples are brachysclereids or the stone cells (called stone cells because of their hardness) of pears (*Pyrus communis*) and quinces (*Cydonia oblonga*) and those of the shoot of the wax plant (*Hoya carnosa*). The cell walls fill nearly all the cell's volume. A layering of the walls and the existence of branched pits is clearly visible. Branched pits such as these are called ramiform pits. The shell of many seeds like those of nuts as well as the stones of drupes like cherries or plums are made up from sclereids. These structures are used to protect other cells.

The Following is a Glossary of Plant Cell Anatomy Terms:

- Amyloplast - an organelle in some plant cells that stores starch. Amyloplasts are found in starchy plants like tubers and fruits.
- ATP - ATP is short for adenosine triphosphate; it is a high-energy molecule used for energy storage by organisms. In plant cells, ATP is produced in the cristae of mitochondria and chloroplasts.
- Cell membrane - the thin layer of protein and fat that surrounds the cell, but is inside the cell wall. The cell membrane is semipermeable, allowing some substances to pass into the cell and blocking others.
- Cell wall - a thick, rigid membrane that surrounds a plant cell. This layer of cellulose fibre gives the cell most of its support and structure. The cell wall also bonds with other cell walls to form the structure of the plant.
- Centrosome - (also called the "microtubule organizing centre") a small body located near the nucleus - it has a dense centre and radiating tubules. The centrosomes is where microtubules are made. During cell division (mitosis), the centrosome divides and the two parts move to opposite sides of the dividing cell. Unlike the centrosomes in animal cells, plant cell centrosomes do not have centrioles.
- Chlorophyll - chlorophyll is a molecule that can use light energy from sunlight to turn water and carbon dioxide gas into sugar and oxygen (this process is called photosynthesis). Chlorophyll is magnesium based and is usually green.
- Chloroplast - an elongated or disc-shaped organelle containing chlorophyll. Photosynthesis (in which energy from sunlight is converted into chemical energy - food) takes place in the chloroplasts.

- Christae - (singular crista) the multiply-folded inner membrane of a cell's mitochondrion that are finger-like projections. The walls of the cristae are the site of the cell's energy production (it is where ATP is generated).
- Cytoplasm - the jellylike material outside the cell nucleus in which the organelles are located.
- Golgi body - (also called the golgi apparatus or golgi complex) a flattened, layered, sac-like organelle that looks like a stack of pancakes and is located near the nucleus. The golgi body packages proteins and carbohydrates into membrane-bound vesicles for "export" from the cell.
- Granum - (plural grana) A stack of thylakoid disks within the chloroplast is called a granum.
- Mitochondrion - spherical to rod-shaped organelles with a double membrane. The inner membrane is infolded many times, forming a series of projections (called cristae). The mitochondrion converts the energy stored in glucose into ATP (adenosine triphosphate) for the cell.
- Nuclear membrane - the membrane that surrounds the nucleus.
- Nucleolus - an organelle within the nucleus - it is where ribosomal RNA is produced.
- Nucleus - spherical body containing many organelles, including the nucleolus. The nucleus controls many of the functions of the cell (by controlling protein synthesis) and contains DNA (in chromosomes). The nucleus is surrounded by the nuclear membrane
- Photosynthesis - a process in which plants convert sunlight, water, and carbon dioxide into food energy (sugars and starches), oxygen and water. Chlorophyll or closely-related pigments (substances that colour the plant) are essential to the photosynthetic process.
- Ribosome - small organelles composed of RNA-rich cytoplasmic granules that are sites of protein synthesis.
- Rough endoplasmic reticulum - (rough ER) a vast system of interconnected, membranous, infolded and convoluted sacks that are located in the cell's cytoplasm (the ER is continuous with the outer nuclear membrane). Rough ER is covered with ribosomes that give it a rough appearance. Rough ER transport

materials through the cell and produces proteins in sacks called cisternae (which are sent to the Golgi body, or inserted into the cell membrane).

- Smooth endoplasmic reticulum - (smooth ER) a vast system of interconnected, membranous, infolded and convoluted tubes that are located in the cell's cytoplasm (the ER is continuous with the outer nuclear membrane). The space within the ER is called the ER lumen. Smooth ER transport materials through the cell. It contains enzymes and produces and digests lipids (fats) and membrane proteins; smooth ER buds off from rough ER, moving the newly-made proteins and lipids to the Golgi body and membranes
- Stroma - part of the chloroplasts in plant cells, located within the inner membrane of chloroplasts, between the grana.
- Thylakoid disk - thylakoid disks are disk-shaped membrane structures in chloroplasts that contain chlorophyll. Chloroplasts are made up of stacks of thylakoid disks; a stack of thylakoid disks is called a granum. Photosynthesis (the production of ATP molecules from sunlight) takes place on thylakoid disks.
- Vacuole - a large, membrane-bound space within a plant cell that is filled with fluid. Most plant cells have a single vacuole that takes up much of the cell. It helps maintain the shape of the cell.

Aerenchyma

Aerenchyma refers to spaces or air channels in the leaves, stems and roots of some plants, which allows exchange of gases between the shoot and the root. The channels of air-filled cavities provide a low-resistance internal pathway for the exchange of gases such as oxygen and ethylene between the plant above the water and the submerged tissues. Aerenchyma is widespread in aquatic and wetland plants which must grow in hypoxic soils.

Aerenchyma Formation and Hypoxia

When soil is flooded, hypoxia develops, as soil microorganisms consume oxygen faster than diffusion occurs. The presence of hypoxic soils is one of the defining characteristics of wetlands. Many wetland plants possess aerenchyma, and in some, such as water-lilies, there is mass flow of atmospheric air through leaves and rhizomes. There are many other chemical consequences of hypoxia. For example, nitrification is inhibited as low oxygen occurs and toxic compounds are formed, as anaerobic bacteria use nitrate, manganese, and sulphate as alternative electron acceptors. The reduction-oxidation potential of the rhizhosphere decreases and metal ions such as iron and manganese precipitate.

In general, low oxygen stimulates trees and plants to produce ethylene. Ethylene slows down primary and adventitious root elongation and formation.

Advantages of Aerenchyma

The large air-filled cavities provide a low-resistance internal pathway for the exchange of gases between the plant organs above the water and the submerged tissues. This allows plants to grow without incurring the metabolic costs of anaerobic respiration. Some of the oxygen transported through the aerenchyma leaks through root pores into the surrounding soil. The resulting small rhizosphere of oxygenated soil around individual roots support microorganisms that prevent the influx of potentially toxic soil components such as sulfide, iron, and manganese.

Acid Growth

Acid growth refers to the ability of plant cells and plant cell walls to elongate or expand quickly at low (acidic) pH. This form of growth does not involve an increase in cell number. During acid growth, plant cells enlarge rapidly because the cell walls are made more extensible by expansin, a pH-dependent wall-loosening protein. Expansin loosens the network-like connections between cellulose microfibrils within the cell wall, which allows the cell volume to increase by turgor and osmosis. A typical sequence leading up to this would involve the introduction of a plant hormone (auxin, for example) that causes protons (H^+ ions) to be pumped out of the cell into the cell wall. As a result the cell wall solution becomes more acidic. This activates expansin activity, causing the wall to become more extensible and to undergo wall stress relaxation, which enables the cell to take up water and to expand.

Amyloplast

Amyloplasts are non-pigmented organelles found in some plant cells. They are responsible for the synthesis and storage of starch granules, through the polymerization of glucose. Amyloplasts also convert this starch back into sugar when the plant needs energy.

Large numbers of amyloplasts can be found in fruit and in underground storage tissues of some plants, such as in potato tubers. Amyloplasts are starch grains.

Amyloplasts are plastids, specifically leucoplasts. Plastids are a specialized class of cellular organelles that carry their own genome and are believed to be descendants of cyanobacteria (blue-green algae) which formed a symbiotic relationship with the eukaryotic cell.

Starch synthesis and storage also takes place in chloroplasts, a type of pigmented plastid involved in photosynthesis. Amyloplasts and chloroplasts are closely related, and amyloplasts can turn into chloroplasts; this is for instance observed when potato tubers are exposed to light and turn green.

Statoliths: Sensing Gravity

In the root cap (a tissue at the tip of the root) there is a special subset of cells, called statocytes. Inside the statocyte cells, some specialized amyloplasts are involved in the perception of gravity by the plant (gravitropism). These specialized amyloplasts—called statoliths—are denser than the cytoplasm and can sediment according to the gravity vector. The statoliths are enmeshed in a web of actin and it is thought that their sedimentation transmits the gravitropic signal by activating mechanosensitive channels. The gravitropic signal then leads to reorientation of auxin efflux carriers and subsequent redistribution of auxin streams in root cap and root as a whole. The changed relations in concentration of auxin leads to differential growth of the root tissues. Taken together, the root is then turning, following the gravity stimuli. They are also found in the endodermic layer of the inflorescence stem. The redistribution of auxin causes the shoot to turn in a direction opposite that of the gravity stimuli.

Apical Cell

An apical cell is a cell located at the tip of a bryophyte gametophyte stem that constantly divides leaving new cells behind to bui.e stem.

Cytokinesis in terrestrial plants occurs by cell plate formation. This process entails the delivery of Golgi-derived and endosomal vesicles carrying cell wall and cell membrane components to the plane of cell division and the subsequent fusion of these vesicles within this plate. After formation of an early tubulo-vesicular network at the centre of the cell, the initially labile cell plate consolidates into a tubular network and eventually a fenestrated sheet. The cell plate grows outward from the centre of the cell to the parental plasma membrane with which it will fuse, thus completing cell division. Formation and growth of the cell plate is dependent upon the phragmoplast, which is required for proper targeting of Golgi-derived vesicles to the cell plate.

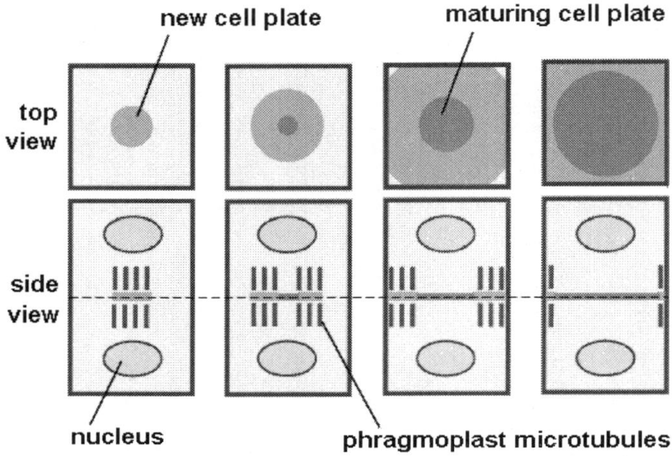

Figure: *Phragmoplast and cell plate formation in a plant cell during cytokinesis. Left side: Phragmoplast forms and cell plate starts to assemble in the centre of the cell. Toawards the right: Phragmoplast enlarges in a donut-shape towards the outside of the cell, leaving behind mature cell plate in the centre. The cell plate will transform into the new cell wall once cytokinesis is complete.*

As the cell plate matures in the central part of the cell, the phragmoplast disassembles in this region and new elements are added on its outside. This process leads to a steady expansion of the phragmoplast and, concomitantly, to a continuous retargeting of Golgi-derived vesicles to the growing edge of the cell plate. Once the cell plate reaches and fuses with the plasma membrane the phragmoplast disappears. This event not only marks the separation of the two daughter cells, but also initiates a range of biochemical modifications that transform the callose-rich, flexible cell plate into a cellulose-rich, stiff primary cell wall.

The heavy dependence of cell plate formation on active Golgi stacks explains why plant cells, unlike mammalian cells, do not disassemble their secretion machinery during cell division.

Elaioplast

Elaioplasts are a type of leucoplast that is specialized for the storage of lipids in plants. Elaioplasts house oil body deposits as rounded plastoglobuli, which are essentially fat droplets.

Being a variety of leucoplast, elaioplasts are non-pigmented and fall into the much broader organelle category of plant plastids. A different example of a storage-specialized leucoplast is the amyloplast, which stores starch.

Guard Cell

Guard cells are specialized cells located in the leaf epidermis of plants. Pairs of guard cells surround tiny stomatal airway pores. These tiny pores in the leaf surface are necessary for gas exchange into and out of the plant; carbon dioxide (CO_2) enters the plant allowing the carbon fixation reactions of photosynthesis to occur. Oxygen (O_2) exits the plant as a byproduct of photosynthesis. The opening and closing of the stomatal gas exchange holes is regulated by swelling and shrinking of the two surrounding guard cells. Water evaporates through the stomatal openings causing plants to lose water. Over 95% of water loss from plants can occur by evaporation (transpiration) through the stomatal pores. Therefore, it is important for plants to be able to balance the amount of CO_2 being brought into the plant with the amount of water escaping as a result of the open stomatal pores. Hence, the guard cells are the gate keepers of the plants ability to take in CO_2 from the atmosphere for photosynthesis – while regulating how much water plants lose to the atmosphere. Opening and closure of the stomatal pore is mediated by changes in the turgor pressure of the two guard cells. The turgor pressure of guard cells is controlled by movements of large quantities of ions and sugars into and out of the guard cells. When guard cells take up these solutes, the water potential (Ø) inside the cells decreases, causing osmotic water flow into the guard cells. This leads to a turgor pressure increase causing swelling of the guard cells and the stomatal pores open. The ions that are taken up by guard cells are mainly potassium (K^+) ions and chloride (Cl^-) ions. In addition guard cells take up sugars that also contribute to opening of the stomatal pores.

Water Loss and Water Use Efficiency

Water stress (drought and salt stress) is one of the major environmental problems causing severe losses in agriculture and in nature. Drought tolerance of plants is mediated by several mechanisms that work together, including stabilizing and protecting the plant from damage caused by desiccation and also controlling how much water plants lose through the stomatal pores during drought. A plant hormone, abscisic acid (ABA), is produced in response to drought. A major type of ABA receptor has been identified. Future research is needed to test if these receptors can be used to engineer drought tolerance in plants. The plant hormone ABA causes the stomatal pores to close in response to drought, which reduces plant water loss via transpiration to the atmosphere and allows plants to avoid or slow down water loss during droughts. The use of drought tolerant crop plants would lead to a reduction in crop losses during droughts. Since guard cells control water loss of plants, the investigation on how stomatal opening and closure are regulated could lead to the development of plants with improved avoidance or slowing of desiccation and better water use efficiency.

Ion Uptake and Release

Ion uptake into guard cells causes stomatal opening: The opening of gas exchange pores requires the uptake of potassium ions into guard cells. Potassium channels and pumps have been identified and shown to function in the uptake of ions and opening of stomatal apertures. Ion release from guard cells causes stomatal pore closing: Other ion channels have been identified that mediate release of ions from guard cells, which results in osmotic water efflux from guard cells due to osmosis, shrinking of the guard cells, and closing of stomatal pores. Specialized potassium efflux channels participate in mediating release of potassium from guard cells. Anion channels were identified as important controllers of stomatal closing. Anion channels have several major functions in controlling stomatal closing: (a) They allow release of anions, such as chloride and malate from guard cells, which is needed for stomatal closing. (b) Anion channels are activated by signals that cause stomatal closing, for example by intracellular calcium and ABA. The resulting release of negatively charged anions from guard cells results in an electrical shift of the membrane to more positive voltages (depolarization) at the intracellular surface of the guard cell plasma membrane. This electrical depolarization of guard cells leads to activation of the outward potassium channels and the release of potassium through these channels. At least two major types of anion channels have been characterized in the plasma membrane: S-type anion channels and R-type anion channels.

Vacuolar Ion Transport

Vacuoles are large intracellular storage organelles in plants cells. In addition to the ion channels in the plasma membrane, vacuolar ion channels have important functions in regulation of stomatal opening and closure because vacuoles can occupy up to 90% of guard cell's volume. Therefore, a majority of ions are released from vacuoles when stomata are closed. Vacuolar K^+ (VK) channels and fast vacuolar channels can mediate K^+ release from vacuoles. Vacuolar K^+ (VK) channels are activated by elevation in the intracellular calcium concentration. Another type of calcium-activated channel, is the slow vacuolar (SV) channel. SV channels have been shown to function as cation channels that are permeable to Ca^{2+} ions, but their exact functions are not yet known in plants.

Signal Transduction

Guard cells perceive and process environmental and endogenous stimuli such as light, humidity, CO_2, temperature, drought, and plant hormones to trigger cellular responses resulting in stomatal opening or closure. These signal transduction pathways determine for example how quickly a plant will lose water during a drought period. Guard cells have become a model for single cell signalling. Using Arabidopsis thaliana, the investigation of signal processing in single guard cells has become open to the power of genetics. Cytosolic and nuclear proteins and chemical messengers that function in stomatal movements have been identified that mediate the transduction of environmental signals thus controlling CO_2 intake into plants and plant water loss.

Research on guard cell signal transduction mechanisms is producing an understanding of how plants can improve their response to drought stress by reducing plant water loss. Guard cells also provide an excellent model for basic studies on how a cell integrates numerous kinds of input signals to produce a response (stomatal opening or closing). These responses require coordination of numerous cell biological processes in guard cells, including signal reception, ion channel and pump regulation, membrane trafficking, transcription, cytoskeletal rearrangements and more. A challenge for future research is to assign the functions of some of the identified proteins to these diverse cell biological processes.

Development

During the development of plant leaves, the specialized guard cells differentiate from "guard mother cells". The density of the stomatal pores in leaves is regulated by environmental signals, including the continuingly increasing atmospheric CO_2 concentration, which reduces the density of stomatal pores in the surface of leaves in many plant species by presently unknown mechanisms. The genetics of stomatal development can be directly studied by imaging of the leaf epidermis using a microscope. Several major control proteins that function in a pathway mediating the development of guard cells and the stomatal pores have been identified.

Leucoplast

Leucoplasts are a category of plastid and as such are organelles found in plant cells. They are non-pigmented, in contrast to other plastids such as the chloroplast.

Lacking pigments, leucoplasts are not green. They are colourless, so they are predictably located in roots and non-photosynthetic tissues of plants. They may become specialized for bulk storage of starch, lipid or protein and are then known as amyloplasts, elaioplasts, or proteinoplasts respectively. However, in many cell types, leucoplasts do not have a major storage function and are present to provide a wide range of essential biosynthetic functions, including the synthesis of fatty acids, many amino acids, and tetrapyrrole compounds such as haem. In general, leucoplasts are much smaller than chloroplasts and have a variable morphology, often described as amoeboid. Extensive networks of stromules interconnecting leucoplasts have been observed in epidermal cells of roots, hypocotyls, and petals, and in callus and suspension culture cells of tobacco. In some cell types at certain stages of development, leucoplasts are clustered around the nucleus with stromules extending to the cell periphery, as observed for proplastids in the root meristem.

Etioplasts, which are pre-granal, immature chloroplasts but can also be chloroplasts that have been deprived of light, lack active pigment and can be considered leucoplasts. After several minutes exposure to light, etioplasts begin to transform into functioning chloroplasts and cease being leucoplasts.

2

Megaspore Mother Cell

A megaspore mother cell, or megasporocyte, is a diploid cell in plants in which meiosis will occur, resulting in the production of four megaspores. In angiosperms or flowering plants, the megasporocyte produces a megaspore that develops into an embryo sac through two distinct processes – megasporogenesis (formation of the megaspore in the nucellus, or megasporangium), and megagametogenesis (development of the megaspore into the embryo sac, or megagametophyte).

Generally, within each sporangium, meiosis of a megaspore mother cell produces four haploid megaspores. In gymnosperms and angiosperms, only one of these four megaspores is functional at maturity, and the other three soon degenerate. The megaspore that remains divides mitotically and develops into the female gametophyte (megagametophyte) which eventually produces one or more eggs.

Oleosin

Oleosins are structural proteins found in vascular plant oil bodies and found in plant cells. Oil bodies are not considered organelles because they have a single layer membrane and lack the pre-requisite double layer membrane in order to be considered an organelle. They are found in plant parts with high oil content that undergo extreme desiccation as part of their maturation process, and help stabilize the bodies.

Oleosins are proteins of 16 Kd to 24 Kd and are composed of three domains: an N-terminal hydrophilic region of variable length (from 30 to 60 residues); a central hydrophobic domain of about 70 residues and a C-terminal amphipathic region of variable length (from 60 to 100 residues). The central hydrophobic domain is proposed to be made up of beta-strand structure and to interact with the lipids. It is the only domain whose sequence is conserved. Models show oleosins having a hairpin-like hydrophobic shape that is inserted inside the triacylglyceride (TAG), while the hydrophilic parts are left outside oil bodies.

Oleosins have been found on oil bodies of seeds, tapetum cells, and pollen but not fruits. Instead of a stabilizer of oil bodies, oleosins are believe to be involved in water-uptaking of pollen on stigma.

Use in Purification of Recombinant Protein

Oleosins provide an easy way of purifying proteins which have been produced recombinantly in plants. If the protein is made as a fusion protein with oleosin and a protease recognition site is incorporated between them, the fusion protein will sit in the membrane of the oil body, which can be easily isolated by centrifugation. The oil droplets can then be mixed with aqueous medium again, and oleosin cleaved from the protein of interest. Centrifugation will cause two phases to separate again, and the aqueous medium now contains the purified protein.

Onion Epidermal Cell

The onion epidermal cell provides a protective layer against unwanted viruses that may harm the sensitive tissues. Because of its simple structure and transparency it is often used to introduce students to plant anatomy. These properties also make it useful for demonstrations of plasmolysis. The clear epidermal cells consist of a single layer and contain chloroplasts, because the onion fruiting body (bulb) is used for storing energy, not photosynthesis. To prepare onion epidermal cells for microscopy, smaller onions that feel firm are best. Remove the epidermal layers by cutting the onion and peeling them off (they are the membrane-like sheaths between each onion layer). For advanced microscopy, such as fluorescence microscopy, the layers halfway between the outside and the centre of the onion are best.

If you are using these layers for biolistics, it is important to note that bombarding with the cells facing up or down, on the media that they're placed on, can't make a difference.

Palisade Cell

Palisade cells are cells found within the mesophyll in leaves of dicotyledonous plants right below the upper epidermis and cuticle. The Palisade layer is on the top side of the leaf to collect as much sunlight as possible. They contain the largest number of chloroplasts per cell, which makes them the primary site of photosynthesis in the leaves of those plants that contain them, converting the energy in light to the chemical energy of carbohydrates. Their cylindrical shape allows a large amount of light to be absorbed by the chloroplasts. These cells are positioned near the top surfaces of the leaves, underneath the epidermis.

Beneath the palisade mesophyll are the spongy mesophyll cells, which also perform photosynthesis. They are irregularly-shaped cells that having many intercellular spaces that allow the passage of gases, such as the carbon dioxide needed for photosynthesis to take place.

Pavement Cells

Pavement cells are epidermal cells that serve to protect the plant body from the dry atmosphere and UV-rays. As a part of the epidermis it is important for protection as these pavement cells secrete waxes to form the cuticle.

Phragmoplast

The phragmoplast is a plant cell specific structure that forms during late cytokinesis. It serves as a scaffold for cell plate assembly and subsequent formation of a new cell wall separating the two daughter cells.

The phragmoplast is a complex assembly of microtubules (MTs), microfilaments (MFs), and endoplasmic reticulum (ER) elements, that assemble in two opposing sets perpendicular to the plane of the future cell plate during anaphase and telophase. It is initially barrel-shaped and forms from the mitotic spindle between the two daughter nuclei while nuclear envelopes reassemble around them. The cell plate initially forms as a disc between the two halves of the phragmoplast structure. While new cell plate material is added to the edges of the growing plate, the phragmoplast microtubules disappear in the centre and regenerate at the edges of the growing cell plate. The two structures grow outwards until they reach the outer wall of the dividing cell. If a phragmosome was present in the cell, the phragmoplast and cell plate will grow through the space occupied by the phragmosome. They will reach the parent cell wall exactly at the position formerly occupied by the preprophase band.

The microtubules and actin filaments within the phragmoplast serve to guide vesicles with cell wall material to the growing cell plate. Actin filaments are also possibly involved in guiding the phragmoplast to the site of the former preprophase band location at the parent cell wall. While the cell plate is growing, segments of smooth endoplasmic reticulum are trapped within it, later forming the plasmodesmata connecting the two daughter cells.

The phragmoplast can only be observed in Embryophytes, that is the bryophytes and vascular plants and a few advanced green algae, specifically *Coleochaete* in the Division Charophyta. Some algae use another type of microtubule array, a phycoplast, during cytokinesis.

Phragmosome

The phragmosome is a sheet of cytoplasm forming in highly vacuolated plant cells in preparation for mitosis. In contrast to animal cells, plant cells often contain large central vacuoles occupying up to 90% of the total cell volume and pushing the nucleus against the cell wall. In order for mitosis to occur, the nucleus has to move into the centre of the cell. This happens during G2 phase of the cell cycle just after DNA replication.

Initially, cytoplasmic strands form that penetrate the central vacuole and provide pathways for nuclear migration. Actin filaments along these cytoplasmic strands pull the nucleus into the centre of the cell. These cytoplasmic strands fuse into a transverse sheet of cytoplasm along the plane of future cell division, forming the phragmosome. Phragmosome formation is only clearly visible in dividing plant cells that are highly vacuolated.

Just before mitosis, a dense band of microtubules appears around the phragmosome and the future division plane just below the plasma membrane. This preprophase band marks the equatorial plane of the future mitotic spindle as well as the future fusion sites for the new cell plate with the existing cell wall. It disappears as soon as the nuclear envelope breaks down and the mitotic spindle forms.

When mitosis is completed, the cell plate and new cell wall form starting from the centre along the plane occupied by the phragmosome. The cell plate grows outwards until it fuses with the cell wall of the dividing cell at exactly the spots predicted by the preprophase band.

Phycoplast

The phycoplast is a microtubule structure observed during cytokinesis in members of the *Chlorophyceae*, the largest class of green algae. Cytokinesis in green algae occurs via a diverse range of mechanisms, including cleavage furrows in some algae and cell plates in others. Some green algae of the class *Charophyceae* use phragmoplasts similar to those in embryophytes to organize and guide the growing cell plate. In these algae, the microtubules of the telophase spindle give rise to the phragmoplast and are oriented perpendicular to the plane of cell division and the forming cell plate. The growth of the cell plate eventually disrupts the telophase spindle.

In the *Chlorophyceae*, the most common form of cell division occurs via a phycoplast. In these algae, the spindle collapses and a new system of microtubules forms that is oriented in parallel to the plane of cell division. This phycoplast can be observed in algae undergoing cytokinesis via cleavage furrow (case 1 in picture) as well as algae utilizing a cell plate (case 3 in picture). The phycoplast may play a role in assuring that the plane of cell division will pass between the two daughter nuclei. Typically, these algae undergo "closed" mitosis where the nuclear envelope persists throughout mitosis.

Preprophase

Preprophase is an additional phase during mitosis in plant cells that does not occur in other eukaryotes such as animals or fungi. It precedes prophase and is characterized by two distinct events:

- The formation of the preprophase band, a dense microtubule ring underneath the plasma membrane.
- The initiation of microtubule nucleation at the nuclear envelope.

Function of Preprophase in the Cell Cycle

Plant cells are fixed with regards to their neighbour cells within the tissues they are growing in. In contrast to animals where certain cells can migrate within the embryo to form new tissues, the seedlings of higher plants grow entirely based on the orientation of cell division and subsequent elongation and differentiation of cells within their cell walls. Therefore, the accurate control of cell division planes and placement of the future cell wall in plant cells is crucial for the correct architecture of plant tissues and organs.

The preprophase stage of somatic plant cell mitosis serves to establish the precise location of the division plane and future cell wall before the cell enters prophase. This is achieved through the formation of a transient microtubule structure, the preprophase band, and a so far unknown mechanism by which the cell is able to "memorize" the position of the preprophase band to guide the new cell wall growing during cytokinesis to the correct location. In gametophyte tissues during the reproductive phase of the plant life cycle, cell division planes may be established without the use of a preprophase band.

In highly vacuolated plant cells, preprophase may be preceded by the formation of a phragmosome. The function of the phragmosome is to suspend the cell nucleus in the centre of the cell in preparation for mitosis. If a phragmosome is visible, the preprophase band will appear at its outer edge.

Preprophase Band Formation

At the beginning of preprophase, the cortical microtubules of a plant cell disappear and aggregate into a dense ring underneath the plasma membrane. This preprophase band runs around the equatorial plane of the future mitotic spindle and marks the plane of cell division and future fusion site for the cell plate. It consists of microtubules and microfilaments (actin) and persists into prophase. Spindle formation occurs during prophase with the axis perpendicular to the plane surrounded by the preprophase band.

Microtubule Nucleation

In contrast to animal cells, plant cells do not possess centrosomes to organize their mitotic spindles. Instead, the nuclear envelope acts as a microtubule organizing centre (MTOC) for spindle formation during preprophase. The first sign is a clear, actin-free zone appearing around the nuclear envelope. This zone fills with microtubules nucleating on the surface of the nucleus. The preprophase spindle forms by self-assembly of these microtubules in the cytoplasm surrounding the nuclear envelope. It is reinforced through chromosome (kinetochore)-mediated spindle assembly after the nuclear envelope breaks down at the end of prophase.

Transition Into Prophase

During progression from preprophase into prophase, the randomly oriented microtubules align parallel along the nuclear surface according to the spindle axis. This structure is called the *prophase spindle.* Triggered by nuclear membrane breakdown at the beginning of prometaphase, the preprophase band disappears and the prophase spindle matures into the metaphase spindle occupying the space of the former nucleus. Experiments with drugs destroying microfilaments indicate that actin may play a role in keeping the cellular "memory" of the position of the division plane after the preprophase band breaks down to direct cytokinesis in telophase.

Preprophase Band

The preprophase band is a microtubule array found in plant cells that are about to undergo cell division and enter the preprophase stage of the plant cell cycle. Besides the phragmosome, it is the first microscopically visible sign that a plant cell is about to enter mitosis. The preprophase band was first observed and described by Jeremy Pickett-Heaps and Donald Northcote at Cambridge University in 1966.

Just before mitosis starts, the preprophase band forms as a dense band of microtubules around the phragmosome and the future division plane just below the plasma membrane. It encircles the nucleus at the equatorial plane of the future mitotic spindle when dividing cells enter the G2 phase of the cell cycle after DNA replication is complete. The preprophase band consists mainly of microtubules and microfilaments (actin) and is generally 2-3 μm wide. When stained with fluorescent markers, it can be seen as two bright spots close to the cell wall on either side of the nucleus.

Plant cells lack centrosomes as microtubule organizing centres. Instead, the microtubules of the mitotic spindle aggregate on the nuclear surface and are reoriented to form the spindle at the end of prophase. The preprophase band also functions in properly orienting the mitotic spindle, and contributes to efficient spindle formation during prometaphase

The preprophase band disappears as soon as the nuclear envelope breaks down and the mitotic spindle forms, leaving behind an actin-depleted zone. However, its position marks the future fusion sites for the new cell plate with the existing cell wall during telophase. When mitosis is completed, the cell plate and new cell wall form starting from the centre along the plane occupied by the phragmosome. The cell plate grows outwards until it fuses with the cell wall of the dividing cell at exactly the spots predicted by the position of the preprophase band.

Statocyte

Statocytes are cells thought to be involved in gravitropic perception in plants, located in the cap tissue of the roots. They contain statoliths – starch-filled amyloplastic organelles – which sediment at the lowest part of the cells and initiate differential growth patterns, bending the root towards the vertical axis.

Stoma

In botany, a stoma (plural *stomata*) (occasionally called a stomate, plural stomates) is a pore, found in the epidermis of leaves, stems and other organs that is used to control gas exchange. The pore is bordered by a pair of specialized parenchyma cells known as guard cells that are responsible for regulating the size of the opening. The term is also used collectively to refer to an entire stomatal complex, both the pore itself and its accompanying guard cells. Air containing carbon dioxide and oxygen enters the plant through these openings and is used in photosynthesis in the mesophyll cells (parenchyma cells with chloroplasts) and respiration, respectively. Oxygen produced as a by-product of photosynthesis diffuses out to the atmosphere through these same openings. Also, water vapor is released into the atmosphere through these pores in a process called transpiration.

Stomata are present in the sporophyte generation of all land plant groups except liverworts. Dicotyledons usually have more stomata on the lower epidermis than the upper epidermis. Monocotyledons, on the other hand, usually have the same number of stomata on the two epidermes. In plants with floating leaves, stomata may be found only on the upper epidermis; submerged leaves may lack stomata entirely.

Function

CO_2 Gain and Water Loss: Carbon dioxide, a key reactant in photosynthesis, is present in the atmosphere at a concentration of about 390 ppm (as of December 2011). Most plants require the stomata to be open during daytime. The problem is that the air spaces in the leaf are saturated with water vapour, which exits the leaf through the stomata (this is known as transpiration). Therefore, plants cannot gain carbon dioxide without simultaneously losing water vapour.

Alternative Approaches

Ordinarily, carbon dioxide is fixed to ribulose-1,5-bisphosphate (RuBP) by the enzyme RuBisCO in mesophyll cells exposed directly to the air spaces inside the leaf. This exacerbates the transpiration problem for two reasons: first, RuBisCo has a relatively low affinity for carbon dioxide, and second, it fixes oxygen to RuBP, wasting energy and carbon in a process called photorespiration. For both of these reasons, RuBisCo needs high carbon dioxide concentrations, which means wide stomatal apertures and, as a consequence, high water loss.

Narrower stomatal apertures can be used in conjunction with an intermediary molecule with a high carbon dioxide affinity, PEPcase (Phosphoenolpyruvate carboxylase). Retrieving the products of carbon fixation from PEPCase is in an energy-intensive process, however. As a result, the PEPCase alternative is preferable only where water is limiting but light is plentiful, or where high temperatures increase the solubility of oxygen relative to that of carbon dioxide, magnifying RuBisCo's oxygenation problem.

CAM Plants

A group of mostly desert plants called "CAM" plants (Crassulacean acid metabolism, after the family Crassulaceae, which includes the species in which the CAM process was first discovered) open their stomata at night (when water evaporates more slowly from leaves for a given degree of stomatal opening), use PEPcarboxylase to fix carbon dioxide and store the products in large vacuoles. The following day, they close their stomata and release the carbon dioxide fixed the previous night into the presence of RuBisCO. This saturates RuBisCO with carbon dioxide, allowing minimal photorespiration. This approach, however, is severely limited by the capacity to store fixed carbon in the vacuoles, so it is preferable only when water is severely limiting.

Opening and Closure

However, most plants do not have the aforementioned facility and must therefore open and close their stomata during the daytime in response to changing conditions, such as light intensity, humidity, and carbon dioxide concentration. It is not entirely certain how these responses work. However, the basic mechanism involves regulation of osmotic pressure.

When conditions are conducive to stomatal opening (e.g., high light intensity and high humidity), a proton pump drives protons (H^+) from the guard cells. This means that the cells' electrical potential becomes increasingly negative. The negative potential opens potassium voltage-gated channels and so an uptake of potassium ions (K^+) occurs. To maintain this internal negative voltage so that entry of potassium ions does not stop, negative ions balance the influx of potassium. In some cases, chloride ions enter, while in other plants the organic ion malate is produced in guard cells. This increase in solute concentration lowers the water potential inside the cell, which results in the diffusion of water into the cell through osmosis. This increases the cell's volume and turgor pressure. Then, because of rings of cellulose microfibrils that prevent the width of the guard cells from swelling, and thus only allow the extra turgor pressure to elongate the guard cells, whose ends are held firmly in place by surrounding epidermal cells, the two guard cells lengthen by bowing apart from one another, creating an open pore through which gas can move.

When the roots begin to sense a water shortage in the soil, abscisic acid (ABA) is released. ABA binds to receptor proteins in the guard cells' plasma membrane and cytosol, which first raises the pH of the cytosol of the cells and cause the concentration of free Ca^{2+} to increase in the cytosol due to influx from outside the cell and release of Ca^{2+} from internal stores such as the endoplasmic reticulum and vacuoles. This causes the chloride (Cl^-) and inorganic ions to exit the cells. Second, this stops the uptake of any further K^+ into the cells and, subsequently, the loss of K^+. The loss of these solutes causes an increase in water potential, which results in the diffusion of water back out of the cell by osmosis. This makes the cell plasmolysed, which results in the closing of the stomatal pores.

Guard cells have more chloroplasts than the other epidermal cells from which guard cells are derived. Their function is controversial.

Inferring Stomatal Behaviour from Gas Exchange

The degree of stomatal resistance can be determined by measuring leaf gas exchange of a leaf. The transpiration rate is dependent on the diffusion resistance provided by the stomatal pores, and also on the humidity gradient between the leaf's internal air spaces and the outside air. Stomatal resistance (or its inverse, stomatal conductance) can therefore be calculated from the transpiration rate and humidity gradient. This allows scientists to investigate how stomata respond to changes in environmental conditions, such as light intensity and concentrations of gases such as water vapor, carbon dioxide, and ozone. Evaporation (*E*) can be calculated as;

$$E = (e_i - e_a) / Pr$$

where e_i and e_a are the partial pressures of water in the leaf and in the ambient air, respectively, *P* is atmospheric prèssure, and *r* is stomatal resistance. The inverse of *r* is conductance to water vapor (*g*), so the equation can be rearranged to;

$$E = (e_i - e_a)g / P$$

and solved for *g*;

$$g = EP / (e_i - e_a)$$

Photosynthetic CO_2 assimilation (*A*) can be calculated from

$$A = (C_a - C_i)g / 1.6P$$

where C_a and C_i are the atmospheric and sub-stomatal partial pressures of CO_2, respectively. The rate of evaporation from a leaf can be determined using a photosynthesis system. These scientific instruments measure the amount of water vapour leaving the leaf and the vapor pressure of the ambient air. Photosynthetic systems may calculate water use efficiency (A/E), g, intrinsic water use efficiency (A/g), and C_i. These scientific instruments are commonly used by plant physiologists to measure CO_2 uptake and thus measure photosynthetic rate.

Evolution

The fossil record has little to say about the evolution of stomata. They may have evolved by the modification of conceptacles from plants' alga-like ancestors. It is clear, however, that the evolution of stomata must have happened at the same time as the waxy cuticle was evolving - these two traits together constituted a major advantage for early terrestrial plants.

Development

There are three major epidermal cell types which all ultimately derive from the L1 tissue layer of the shoot apical meristem, called protodermal cells: trichomes, pavement cells and guard cells, all of which are arranged in a non-random fashion. Production is reliant on interactions between SPCH (speechless), EPF (downregulates stomata), TMM (too many mouths, downregulates stomata) and stomagen (upregulates stomata, inhibits SPCH), ERL and YODA downregulate stomata too. Stomata positioning is down to CO2 activating EPF1, which activates TMM/ERL which together activate YODA, YODA in turn inhibits SPCH, in turn SPCH activation decreases, allowing asymmetry.

An asymmetrical cell division occurs in protodermal cells resulting in one large cell that is fated to become a pavement cell and a smaller cell called a meristemoid that will eventually differentiate into the guard cells that surround a stoma. This meristemoid then divides asymmetrically one to three times before differentiating into a guard mother cell. The guard mother cell then makes one symmetrical division, which forms a pair of guard cells.

Stomata as Pathogenic Pathways

Stomata are an obvious hole in the leaf by which, as was presumed for a while, pathogens can enter unchallenged. However, it has been recently shown that stomata do in fact sense the presence of some, if not all, pathogens. However, with the virulent bacteria applied to *Arabidopsis* plant leaves in the experiment, the bacteria released the chemical coronatine, which forced the stomata open again within a few hours.

Transfer Cell

Transfer cells are specialized parenchyma cells that have an increased surface area, due to infoldings of the plasma membrane. They facilitate the transport of sugars from a sugar source, mainly leaves, to a sugar sink, often developing fruits. They are found in nectaries of flowers and some carnivorous plants. Transfer cells are specially found in plants in the region of absorption or secretion of nutrients.

Vascular Cambium

The vascular cambium (pl. cambia or cambiums) is a lateral meristem in the vascular tissue of plants. It is a cylinder of unspecialized meristematic cells that divide to give rise to cells that further divide, differentiate and specialize to form the secondary vascular tissues. The vascular cambium is the source of both the secondary xylem (inwards, towards the pith) and the secondary phloem (outwards), and is located between these tissues in the stem and root. A few leaf types also have a vascular cambium.

Origin

Vascular cambium arises from the primary meristem, procambium that remains undifferentiated between the primary xylem and primary phloem. Upon maturity, this region is known as the fascicular cambium, and the area of cells between the vascular bundles (fascicles) called pith rays becomes what is called the interfascicular cambium. The fascicular and interfascicular cambiums, therefore, represent a continuous ring which bisects the primary xylem and primary phloem. The vascular cambium then produces secondary xylem on the inside of the ring, and secondary phloem on the outside, pushing the primary xylem and phloem apart.

The vascular cambium usually consists of two types of cells:

- Fusiform initials (tall cells, axially oriented);
- Ray initials (almost isodiametric cells - smaller and round to angular in shape).

The vascular cambium is a type of meristem - tissue consisting of embryonic (incompletely differentiated) cells from which other (more differentiated) plant tissues originate. Primary meristems are the apical meristems on root tips and shoot tips. Another lateral meristem is the cork cambium, which produces cork, part of the bark. Together, the secondary vascular tissues (produced by the vascular cambium) and periderm (formed by the cork cambium) makes up the secondary plant body. Vascular cambia are found in dicots and gymnosperms but not monocots, which usually lack secondary growth. For successful grafting, the vascular cambia of the stock and scion must be aligned so they can grow together. In wood, the vascular cambium is the obvious line separating the bark and wood.

Synonyms

- Wood cambium
- Main cambium
- Bifacial cambium

Xylem

Xylem is one of the two types of transport tissue in vascular plants (phloem is the other). The word *xylem* is derived from the Greek word *xylon,* meaning "wood"; the best-known xylem tissue is wood, though it is found throughout the plant. Its basic function is to transport water, but it also transports some nutrients through the plant.

Structure

Xylem also contains two other cell types: parenchyma and fibres. Xylem can be found:

- in vascular bundles, present in non-woody plants and non-woody parts of woody plants
- in secondary xylem, laid down by a meristem called the vascular cambium in woody plants
- as part of a stelar arrangement not divided into bundles, as in many ferns.

In transitional stages of plants with secondary growth, the first two categories are not mutually exclusive, although usually a vascular bundle will contain *primary xylem* only.

The branching pattern exhibited by xylem follows Murray's law.

Primary and Secondary Xylem

Primary xylem is the xylem that is formed during primary growth from procambium. It includes protoxylem and metaxylem. Metaxylem develops after the protoxylem but before secondary xylem. Metaxylem has wider vessels and tracheids than protoxylem.

Secondary xylem is the xylem that is formed during secondary growth from vascular cambium. Although secondary xylem is also found in members of the "gymnosperm" groups Gnetophyta and

Ginkgophyta and to a lesser extent in members of the Cycadophyta, the two main groups in which secondary xylem can be found are:

1. conifers (*Coniferae*): there are some six hundred species of conifers. All species have secondary xylem, which is relatively uniform in structure throughout this group. Many conifers become tall trees: the secondary xylem of such trees is used and marketed as softwood.
2. angiosperms (*Angiospermae*): there are some quarter of a million to four hundred thousand species of angiosperms. Within this group secondary xylem is rare in the monocots. Many non-monocot angiosperms become trees, and the secondary xylem of these is used and marketed as hardwood.

Main Function – upwards Water Transport

The xylem transports water and soluble mineral nutrients from the roots throughout the plant. It is also used to replace water lost during transpiration and photosynthesis. Xylem sap consists mainly of water and inorganic ions, although it can contain a number of organic chemicals as well.

The transport is passive, not powered by energy spent by the tracheary elements themselves, which are dead by maturity and no longer have living contents. Two phenomena cause xylem sap to flow:

- Transpirational pull: the most important cause of xylem sap flow is the evaporation of water from the surfaces of mesophyll cells to the atmosphere. This causes millions of minute menisci to form in the mesophyll cell wall. The resulting surface tension causes a negative pressure or tension in the xylem that pulls the water from the roots and soil.
- Root pressure: If the water potential of the root cells is more negative than that of the soil, usually due to high concentrations of solute, water can move by osmosis into the root from the soil. This causes a positive pressure that forces sap up the xylem towards the leaves. In some circumstances, the sap will be forced from the leaf through a hydathode in a phenomenon known as guttation.

  Root pressure is highest in the morning before the stomata open and allow transpiration to begin. Different plant species can have different root pressures even in a similar environment; examples include up to 145 kPa in *Vitis riparia* but around zero in *Celastrus orbiculatus*.

The primary force that creates the capillary action movement of water upwards in plants is the adhesion between the water and the surface of the xylem conduits. Capillary action provides the force that establishes an equilibrium configuration, balancing gravity. When transpiration removes water at the top, the flow is needed to return to the equilibrium.

Transpirational pull results from the evaporation of water from the surfaces of cells in the leaves. This evaporation causes the surface of the water to recess into the pores of the cell wall. By capillary action, the water forms concave menisci inside the pores. The high surface tension of water pulls the concavity outwards, generating enough force to lift water as high as a hundred metres from ground level to a tree's highest branches.

Transpirational pull requires that the vessels transporting the water are very small in diameter, otherwise cavitation would break the water column. And as water evaporates from leaves, more is drawn up through the plant to replace it. When the water pressure within the xylem reaches extreme levels due to low water input from the roots (if, for example, the soil is dry), then the gases come out of solution and form a bubble – an embolism forms, which will spread quickly to other adjacent cells, unless bordered pits are present (these have a plug-like structure called a torus, that seals off the opening between adjacent cells and stops the embolism from spreading).

Cohesion-tension Theory

The *cohesion-tension theory* is a theory of intermolecular attraction that explains the process of water flow upwards (against the force of gravity) through the xylem of plants. It was proposed in 1894 by John Joly and Henry Horatio Dixon. Despite numerous objections, this is the most widely accepted theory for the transport of water through a plant's vascular system based on the classical research of Dixon-Joly (1894), Askenasy (1895), and Dixon (1914,1924).

Water is a polar molecule. When two water molecules approach one another, the slightly negatively charged oxygen atom of one forms a hydrogen bond with a slightly positively charged hydrogen atom in the other. This attractive force, along with other intermolecular forces) is one of the principal factors responsible for the occurrence of surface tension in liquid water. It also allows plants to draw water from the root through the xylem to the leaf.

Water is constantly lost through transpiration from the leaf. When one water molecule is lost another is pulled along by the processes of cohesion and tension. Transpiration pull, utilizing capillary action and the inherent surface tension of water, is the primary mechanism of water movement in plants. However, it is not the only mechanism involved. Any use of water in leaves forces water to move into them.

Transpiration in leaves creates tension (differential pressure) in the cell walls of mesophyll cells. Because of this tension, water is literally being pulled up from the roots into the leaves, helped by cohesion (the pull between individual water molecules, due to hydrogen bonds) and adhesion (the stickiness between water molecules and the hydrophilic cell walls of plants). This mechanism of water flow works because of water potential (water flows from high to low potential), and the rules of simple diffusion.

Over the past century, there has been a great deal of research regarding the mechanism of xylem sap transport; today, most plant scientists continue to agree that the *cohesion-tension theory* best explains this process, but multiforce theories that hypothesize several alternative mechanisms have been suggested, including longitudinal cellular and xylem osmotic pressure gradients, axial potential gradients in the vessels, and gel- and gas-bubble-supported interfacial gradients.

Measurement of Pressure

Until recently, the differential pressure (suction) of transpirational pull could only be measured indirectly, by applying external pressure with a pressure bomb to counteract it. When the technology to perform direct measurements with a pressure probe was developed, there was initially some doubt about whether the classic theory was correct, because some workers were unable to demonstrate negative pressures. More recent measurements do tend to validate the classic theory, for the most part. Xylem transport is driven by a combination of transpirational pull from above and root pressure from below, which makes the interpretation of measurements more complicated.

Evolution

Xylem appeared early in the history of terrestrial plant life. Fossil plants with anatomically preserved xylem are known from the Silurian (more than 400 million years ago), and trace fossils resembling individual xylem cells may be found in earlier Ordovician rocks. The earliest true and recognizable xylem consists of tracheids with a helical-annular reinforcing layer added to the cell wall. This is the only type of xylem found in the earliest vascular plants, and this type of cell continues to be found in the *protoxylem* (first-formed xylem) of all living groups of plants. Several groups of plants later developed pitted tracheid cells, it seems, through convergent evolution. In living plants, pitted tracheids do not appear in development until the maturation of the *metaxylem* (following the *protoxylem*).

In most plants, pitted tracheids function as the primary transport cells. The other type of tracheary element, besides the tracheid, is the vessel element. Vessel elements are joined by perforations into vessels. In vessels, water travels by *bulk flow*, as in a pipe, rather than by diffusion through cell membranes. The presence of vessels in xylem has been considered to be one of the key innovations that led to the success of the angiosperms. However, the occurrence of vessel elements is not restricted to angiosperms, and they are absent in some archaic or "basal" lineages of the angiosperms: (e.g., Amborellaceae, Tetracentraceae, Trochodendraceae, and Winteraceae), and their secondary xylem is described by Arthur Cronquist as "primitively vesselless". Cronquist considered the vessels of *Gnetum* to be convergent with those of angiosperms. Whether the absence of vessels in basal angiosperms is a primitive condition is contested, the alternative hypothesis states that vessel elements originated in a precursor to the angiosperms and were subsequently lost.

To photosynthesize, plants must absorb CO_2 from the atmosphere. However, this comes at a price: while stomata are open to allow CO_2 to enter, water can evaporate. Water is lost much faster than CO_2 is absorbed, so plants need to replace it, and have developed systems to transport water from the moist soil to the site of photosynthesis. Early plants sucked water between the walls of their cells, then evolved the ability to control water loss (and CO_2 acquisition) through the use of stomata. Specialized water transport tissues soon evolved in the form of hydroids, tracheids, then secondary xylem, followed by an endodermis and ultimately vessels.

The high CO_2 levels of Silurian-Devonian times, when plants were first colonizing land, meant that the need for water was relatively low. As CO_2 was withdrawn from the atmosphere by plants, more water was lost in its capture, and more elegant transport mechanisms evolved. As water transport mechanisms, and waterproof cuticles, evolved, plants could survive without being continually covered by a film of water. This transition from poikilohydry to homoiohydry opened up new potential for colonization. Plants then needed a robust internal structure that held long narrow channels for transporting water from the soil to all the different parts of the above-soil plant, especially to the parts where photosynthesis occurred.

During the Silurian, CO_2 was readily available, so little water needed expending to acquire it. By the end of the Carboniferous, when CO_2 levels had lowered to something approaching today's, around 17 times more water was lost per unit of CO_2 uptake. However, even in these "easy" early days, water was at a premium, and had to be transported to parts of the plant from the wet soil to avoid desiccation. This early water transport took advantage of the cohesion-tension mechanism inherent in water. Water has a tendency to diffuse to areas that are drier, and this process is accelerated when water can be wicked along a fabric with small spaces. In small passages, such as that between the plant cell walls (or in tracheids), a column of water behaves like rubber – when molecules evaporate from one end, they literally pull the molecules behind them along the channels. Therefore transpiration alone provided the driving force for water transport in early plants. However, without dedicated transport vessels, the cohesion-tension mechanism cannot transport water more than about 2 cm, severely limiting the size of the earliest plants. This process demands a steady supply of water from one end, to maintain the chains; to avoid exhausting it, plants developed a waterproof cuticle. Early cuticle may not have had pores but did not cover the entire plant surface, so that gas exchange could continue. However, dehydration at times was inevitable; early plants cope with this by having a lot of water stored between their cell walls, and when it comes to it sticking out the tough times by putting life "on hold" until more water is supplied.

To be free from the constraints of small size and constant moisture that the parenchymatic transport system inflicted, plants needed a more efficient water transport system. During the early Silurian, they developed specialized cells, which were lignified (or bore similar chemical compounds) to avoid implosion; this process coincided with cell death, allowing their innards to be emptied and water to be passed through them. These wider, dead, empty cells were a million times more conductive than the inter-cell method, giving the potential for transport over longer distances, and higher CO_2 diffusion rates.

The earliest macrofossils to bear water-transport tubes are Silurian plants placed in the genus *Cooksonia*. The early Devonian pretracheophytes *Aglaophyton* and *Horneophyton* have structures very similar to the hydroids of modern mosses. Plants continued to innovate new ways of reducing the resistance to flow within their cells, thereby increasing the efficiency of their water transport. Bands on the walls of tubes, in fact apparent from the early Silurian onwards, are an early improvisation to aid the easy flow of water. Banded tubes, as well as tubes with pitted ornamentation on their walls, were lignified and, when they form single celled conduits, are considered to be tracheids. These, the "next generation" of transport cell design, have a more rigid structure than hydroids, allowing them to cope with higher levels of water pressure. Tracheids may have a single evolutionary origin, possibly within the hornworts, uniting all tracheophytes (but they may have evolved more than once).

Water transport requires regulation, and dynamic control is provided by stomata. By adjusting the amount of gas exchange, they can restrict the amount of water lost through transpiration. This is an important role where water supply is not constant, and indeed stomata appear to have evolved before tracheids, being present in the non-vascular hornworts.

An endodermis probably evolved during the Silu-Devonian, but the first fossil evidence for such a structure is Carboniferous. This structure in the roots covers the water transport tissue and regulates ion exchange (and prevents unwanted pathogens etc. from entering the water transport system). The endodermis can also provide an upwards pressure, forcing water out of the roots when transpiration is not enough of a driver.

Once plants had evolved this level of controlled water transport, they were truly homoiohydric, able to extract water from their environment through root-like organs rather than relying on a film of surface moisture, enabling them to grow to much greater size. As a result of their independence from their surroundings, they lost their ability to survive desiccation – a costly trait to retain.

During the Devonian, maximum xylem diameter increased with time, with the minimum diameter remaining pretty constant. By the middle Devonian, the tracheid diameter of some plant lineages (Zosterophyllophytes) had plateaued. Wider tracheids allow water to be transported faster, but the overall transport rate depends also on the overall cross-sectional area of the xylem bundle itself. The increase in vascular bundle thickness further seems to correlate with the width of plant axes, and plant height; it is also closely related to the appearance of leaves and increased stomatal density, both of which would increase the demand for water.

While wider tracheids with robust walls make it possible to achieve higher water transport pressures, this increases the problem of cavitation. Cavitation occurs when a bubble of air forms within a vessel, breaking the bonds between chains of water molecules and preventing them from pulling more water up with their cohesive tension. A tracheid, once cavitated, cannot have its embolism removed and return to service (except in a few advanced angiosperms which have developed a mechanism of doing so). Therefore it is well worth plants' while to avoid cavitation occurring. For this reason, pits in tracheid walls have very small diameters, to prevent air entering and allowing bubbles to nucleate. Freeze-thaw cycles are a major cause of cavitation. Damage to a tracheid's wall almost inevitably leads to air leaking in and cavitation, hence the importance of many tracheids working in parallel.

Cavitation is hard to avoid, but once it has occurred plants have a range of mechanisms to contain the damage. Small pits link adjacent conduits to allow fluid to flow between them, but not air – although ironically these pits, which prevent the spread of embolisms, are also a major cause of them. These pitted surfaces further reduce the flow of water through the xylem by as much as 30%. Conifers, by the Jurassic, developed an ingenious improvement, using valve-like structures to isolate cavitated elements. These torus-margo structures have a blob floating in the middle of a donut; when one side depressurizes the blob is sucked into the torus and blocks further flow. Other plants simply accept cavitation; for instance, oaks grow a ring of wide vessels at the start of each spring, none of which survive the winter frosts. Maples use root pressure each spring to force sap upwards from the roots, squeezing out any air bubbles.

Growing to height also employed another trait of tracheids – the support offered by their lignified walls. Defunct tracheids were retained to form a strong, woody stem, produced in most instances by a secondary xylem. However, in early plants, tracheids were too mechanically vulnerable, and retained a central position, with a layer of tough sclerenchyma on the outer rim of the stems. Even when tracheids do take a structural role, they are supported by sclerenchymatic tissue.

Tracheids end with walls, which impose a great deal of resistance on flow; vessel members have perforated end walls, and are arranged in series to operate as if they were one continuous vessel. The function of end walls, which were the default state in the Devonian, was probably to avoid embolisms. An embolism is where an air bubble is created in a tracheid. This may happen as a result of freezing, or by gases dissolving out of solution. Once an embolism is formed, it usually cannot be removed; the affected cell cannot pull water up, and is rendered useless.

End walls excluded, the tracheids of prevascular plants were able to operate under the same hydraulic conductivity as those of the first vascular plant, *Cooksonia*.

The size of tracheids is limited as they comprise a single cell; this limits their length, which in turn limits their maximum useful diameter to 80 ìm. Conductivity grows with the fourth power of diameter, so increased diameter has huge rewards; vessel elements, consisting of a number of cells, joined at their ends, overcame this limit and allowed larger tubes to form, reaching diameters of up to 500 ìm, and lengths of up to 10 m.

Vessels first evolved during the dry, low CO_2 periods of the late Permian, in the horsetails, ferns and Selaginellales independently, and later appeared in the mid Cretaceous in angiosperms and gnetophytes. Vessels allow the same cross-sectional area of wood to transport around a hundred times more water than tracheids! This allowed plants to fill more of their stems with structural fibres, and also opened a new niche to vines, which could transport water without being as thick as the tree they grew on. Despite these advantages, tracheid-based wood is a lot lighter, thus cheaper to make, as vessels need to be much more reinforced to avoid cavitation.

Development

Xylem development can be described by four terms: centrarch, exarch, endarch and mesarch. As it develops in young plants, its nature changes from *protoxylem* to *metaxylem* (i.e. from *first xylem* to *after xylem*). The patterns in which protoxylem and metaxylem are arranged is important in the study of plant morphology.

Protoxylem and Metaxylem

As a young vascular plant grows, one or more strands of primary xylem form in its stems and roots. The first xylem to develop is called 'protoxylem'. In appearance protoxylem is usually distinguished by narrower vessels formed of smaller cells. Some of these cells have walls which contain thickenings in the form of rings or helices. Functionally, protoxylem can extend: the cells are able to grow in size and develop while a stem or root is elongating. Later, 'metaxylem' develops in the strands of xylem. Metaxylem vessels and cells are usually larger; the cells have thickenings which are typically either in the form of ladderlike transverse bars (scalariform) or continuous sheets except for holes or pits (pitted). Functionally, metaxylem completes its development after elongation ceases when the cells no longer need to grow in size.

Patterns of Protoxylem and Metaxylem

There are four main patterns to the arrangement of protoxylem and metaxylem in stems and roots.

- *Centrarch* refers to the case in which the primary xylem forms a single cylinder in the centre of the stem and develops from the centre outwards. The protoxylem is thus found in the central core and the metaxylem in a cylinder around it. This pattern was common in early land plants, such as "rhyniophytes".

The other three terms are used where there is more than one strand of primary xylem.

- *Exarch* is used when there is more than one strand of primary xylem in a stem or root, and the xylem develops from the outside inwards towards the centre, i.e. centripetally. The metaxylem is thus closest to the centre of the stem or root and the protoxylem closest to the periphery. The roots of vascular plants are normally considered to have exarch development.
- *Endarch* is used when there is more than one strand of primary xylem in a stem or root, and the xylem develops from the inside outwards towards the periphery, i.e. centrifugally. The protoxylem is thus closest to the centre of the stem or root and the metaxylem closest to the periphery. The stems of seed plants typically have endarch development.
- *Mesarch* is used when there is more than one strand of primary xylem in a stem or root, and the xylem develops from the middle of a strand in both directions. The metaxylem is thus on both the peripheral and central sides of the strand with the protoxylem between the metaxylem (possibly surrounded by it). The leaves and stems of many ferns have mesarch development.

Plant Cell Functions

Ever wondered what are the different parts of a plant cell? and how the plant cell functions? Scroll down to understand the different plant cell types, and parts and functioning of a plant cell.

For every biology students, studying the animal and plant cell is a compulsory topic. And I hope it is a subject of common interest for those who have a passion for learning life's processes. The different parts of a plant cell work in coordination to carry out the normal physiological and biochemical processes of the plant. Thus, we can rather say that plant cell functions are attributed to each and every part of the plant.

Plant Cell: Types and Functions

Plant cells are of three basic types, namely, parenchyma, collenchyma and sclerenchyma. Each of these cells have different structure and perform varied roles in the overall functioning of the plant. The structure and function of these three plant cells are:

Parenchyma Cells: The parenchyma cells are living and perform multiple functions including storage, light harvesting and gas exchange. Some of them have the ability to differentiate into different types of cells (totipotent) as per the requirements of the plant. Parenchyma cells are present in the leaves, tubers and seeds.

Collenchyma Cells: Collenchyma cells are living only during maturity and possess two walls (primary and secondary). Initially, they are similar to parenchyma cells, which later differentiates into specific collenchyma cells. They are present in the vascular bundles and stems, and aid in supporting the plant during the active growing phase.

Sclerenchyma Cells: Sclerenchyma cells are dead at maturity and have thick secondary walls with lignin deposition. These cells function exclusively as mechanical support of the plant. A typical example of sclerenchyma cells in plants is fibre.

Plant Cell: Parts and Functions

While speaking on plant cell functions, the plant cell organelles play an essential role in carrying out the regular activities of the cell, so that the plant remains alive. For example, photosynthesis which is a characteristic of the plants is performed in the chloroplast; while synthesis of ATP (adenosine triphosphate), a form of energy takes place in the mitochondria. Following is a list of the plant cell parts and their respective functions.

Cell Wall: The outermost covering of the plant cell is the protective layer, the cell wall. Its main function includes giving support, maintaining the cell shape and controlling the growth of the cell.

Cell Membrane: Next to the cell wall, lies the cell membrane that comprises a protein and lipid bilayer. Its main function is selective transport of nutrients, wherein some are allowed to enter the cell, while others are restricted.

Vacuole: Vacuoles are organelles, which shape and structure, alter with respect to the cell requirements. They are filled with a water like solution that contains enzymes, organic and inorganic molecules.

Nucleus: The cell nucleus is simply called the control centre of the plant cell, as it contains hereditary material along with other essential cell components. Overall, the nucleus is responsible for protein synthesis, cell growth, division and development.

Cytoplasm: The portion of the plant cell excluding the nucleus is called cytoplasm. Filled with jelly-like cytoplasmic fluid, maximum organelles are present in the cytoplasm.

Plastids: These are the organelles, which perform the function of photosynthesis and storage of starch molecules. Plastids are of different types and contain photosynthetic pigments.

Mitochondria: Mitochondria, also known as powerhouse of the cell, plays the crucial role of generating chemical energy for proper functioning of the plant cell. They are present in many numbers and contain hereditary material.

Ribosomes: Ribosomes are of two types, attached and free. The former is found attached to the endoplasmic reticulum, while the latter is suspended freely in the cytoplasm. Both ribosomes are responsible for protein synthesis.

Golgi Bodies: Golgi bodies are made up of 4-8 stacks (called cisternae) and are useful for packaging macromolecules that are synthesized by the cell. They are also responsible for transportation of nutrients.

Endoplasmic Reticulum: This is the organelle that connects the nucleus and cytoplasm. It performs the function of synthesizing and storing steroids and glycogen. Endoplasmic reticulum with attached ribosomes are called rough endoplasmic reticulum (RER).

Peroxisomes: These are microbodies of the plant cells that contain various degradation enzymes. Peroxisomes play the major role of digesting complex fatty acids including aiding in photosynthesis.

Thus, a plant cell functions smoothly with the help of its various structural components. Though it is eukaryotic like that of animals, it differs significantly from an animal cell. You may like to know more about the similarities between plant and animal cells. The key distinguishing feature is the presence of cell wall and chloroplast in plant cells, while they are absent in animal cells. If viewed under the microscope, you can see large, prominent vacuoles in the centre of a plant cell, whereas an animal cell has small, inconspicuous vacuoles. With this, I hope you got the answer for your queries regarding plant cell parts and functions.

Similarities Between Plant and Animal Cells

Both animal and plant cells are eukaryotic cells and have several similarities. The similarities include common organelles like cell membrane, cell nucleus, mitochondria, endoplasmic reticulum, ribosomes and golgi apparatus.

All living things are made up of units called cells. A cell, the basic unit of life, is a microscopic structure that forms the basic structural as well as functional unit of living organisms. Cells can be divided into two categories called prokaryotic and eukaryotic cells.

Prokaryotic cells are simple cells that lack nucleus and membrane-bound organelles whereas, eukaryotic cells are more complex cells comprising nuclear and cytoplasmic material sheathed by a cell membrane. Both plants and animals come under eukaryotic cells. Eukaryotic cells have a standard set of features, however, plant and animal cells do have their differences.

Cell Organelles	*Plant Cell Diagram*	*Animal Cell Diagram*
Cell Wall	Present	Absent
Cell Membrane	Present	Present
Cell Nucleus	Present	Present
Mitochondria	Present	Present
Ribosomes	Present	Present
Endoplasmic Reticulum	Present	Present
Golgi Apparatus	Present	Present
Chloroplast	Present	Absent
Vacuole	Present	Present
Centrioles	Absent	Present
Lysosomes	Absent	Present

From the table, we find that besides cell wall, chloroplasts, lysosomes and centrioles, plant and animal cells have the same organelles. Let's learn more about these similarities.

Cell Membrane

Also known as plasma membrane, this membrane is the outermost limiting membrane of the cell that encompasses all the cell contents. This membrane is formed from proteins and lipids and acts as an interface between the cell organelles dunked in the cytoplasm inside the cell and the extracellular fluid on the cell's exterior, which bathes all the cells. It is a semi-permeable membrane and permits the passage of selective substances from the exterior to the interior of the cell and vice versa. Besides this main membrane, the cell also features elaborate network of internal membranes which envelop the different cell organelles, forming several membrane-enclosed compartments within the cell.

Cell Nucleus

The hallmark of a eukaryotic cell is the presence of a membrane-bound nucleus. The very term 'eukaryotic' means possessing a 'true nucleus'. Plants and animal cells have a nucleus, which is a spherical body containing several organelles, nucleolus and chromosomes consisting of DNA. The nucleus is enveloped by a nuclear membrane, which keeps the contents of the nucleus from dissipating into the cytoplasm of the cell. The nucleus controls various functions of the cell by controlling protein synthesis.

Mitochondria

These are cell organelles enclosed by membranes and are scattered in the cytoplasm of the cell. Mitochondria features an outer membrane that encloses the entire structure and an inner membrane that encloses a fluid-filled matrix. The inner membrane features shelf-like cristae projecting into the matrix. It also possesses some 5-10 circular molecules of DNA. Mitochondria are cells responsible for production of energy in the form of ATP. Thus, they are also known as the powerhouse of the cell. Their numbers vary from a few hundreds to a few thousands in a cell, depending on how active the cell is. Mitochondrial cells can increase their number by a process of fission and can reduce their numbers by fusing together.

Endoplasmic Reticulum (ER)

It is an extensive membranous labyrinth, which accounts for half the total membrane in eukaryotic cells. The term 'endoplasmic' means within the cytoplasm and 'reticulum' refers to network. ER consists of a network of membranous sacs and tubules called cisternae. There are two types of ER - rough ER and smooth ER. Rough ER is the endoplasmic reticulum that has a bumpy appearance due to the presence of ribosomes attached to them. The smooth ER does not have ribosomes adhering to them. Rough ER is responsible for protein synthesis, while the smooth ER synthesizes lipids and is responsible for carbohydrate metabolism and detoxification of drugs and other poisons.

Ribosomes

These are small, spherical, non-membranous organelles composed of ribonucleic acid (RNA). Ribosomes comprise 65% ribosomal RNA and 35% ribosomal proteins. They are responsible for protein synthesis and are sites where the cell assembles proteins in accordance to the genetic instructions obtained. Cells which have high protein synthesis rate feature the presence of a large number of ribosomes. Ribosomes are found scattered in the cytoplasm as well as adhering to the outer portion of the endoplasmic reticulum.

Golgi Apparatus

These are stacks of cisternae (flattened sacs of unit membrane) that are mainly devoted to processing the proteins formed in the ER. Their job is to transform proteins into more complex molecules. The Golgi complex also contains a large number of vesicles, which are used to send molecules to the cellular membrane, where excretion is carried out. The Golgi apparatus is mostly responsible for molecular traffic direction in the cell.

Vacuole

Found in all plant cells and most animal cells, vacuoles are fluid-filled sacs present in the cytoplasm of cells, that have no definite shape or size. Their main function is storage; storing gases, fluid, nutrients, poison, waste material, etc. Vacuoles function differently in plant and animal cells. They carry out a large number of different functions right from excretion of wastes, to inter-cellular digestion, maintaining turgor pressure, cell pH, etc. In plant cells, the vacuoles are large and in fully grown plants, only a single large vacuole is seen. However, in animal cells, several small vacuoles are present. Therefore, even though both animal and plant cells have vacuoles, they have a lot of differences.

Besides these similarities, plant and animal cells have a few differences. For example, plant cells include a cell wall that surrounds the cell membrane, thereby making the cells in plants more sturdy. They also contain chloroplasts which capture sunlight for photosynthesis and a large central vacuole for water storage. Animal cells contain organelles called centrioles, responsible for animal cell division, which are not found in plant cells. Moreover, animal cells also possess lysosomes that perform the same role as the vacuole in the plant cell.

Water in Plants

The movement of molecules, specifically water and any solutes, is vital to understand in light of plant processes. This will be more or less a quick review of several guiding principles of water motion in reference to plants.

Molecular Movement

1. Diffusion—Diffusion is the net movement of molecules or ions from an area of higher concentration to an area of lower concentration. Think of it as a rebalancing. The molecules or ions are said to be moving along a diffusion gradient. If molecules or ions moving in the opposite direction are said to be moving against a diffusion gradient. Diffusion will continue until a state of equilibrium is reached. Rates of diffusion are affected by temperature and the density of the involved molecules among other things. In the leaves, water diffuses out via the stomata into the atmosphere.
2. Osmosis—Osmosis in plant cells is basically the diffusion of molecules through a semipermeable, or differentially permeable, membrane from a region of higher solute concentration to a region of lower solute concentration. The application of pressure can prevent osmosis from occurring. Plant physiologists like to describe osmosis more precisely in terms of potentials. Osmotic potential is the minimum pressure required to prevent fluid from moving as a result of osmosis. Fluid will enter the cell via osmosis until the osmotic potential is balanced by the cell wall resistance to expansion. Any water gained by osmosis may help keep a plant cell rigid or turgid. The turgor pressure that develops against the cell walls as a result of water entering the cell's vacuole. This pressure is also referred to as the pressure potential. The crunch when you bite into a celery stick is as a result of the violation of the cell's turgor pressure. The osmotic potential and pressure potential combined make up the water potential of a plant cell. If there are two cells next to each other of different water potentials, water will move from the cell with

the higher water potential to the cell with the lower water potential. Water enters plant cells from the environment via osmosis. Water moves because the overall water potential in the soil is higher than the water potential in the roots and plant parts. If the soil is desiccated then there will be no net movement into the plant cells and the plant will die.

3. Plasmolysis—Plasmolysis is the loss of water via osmosis and accompanying shrinkage of the protoplasm away from the cell wall. When this occurs, the cell is said to be plasmolyzed. This process can be reversed if the cell is placed in fresh water and the cell is allowed to regain its turgor pressure. However, as with anything living, there is a point of no return and permanent or fatal damage to the cell can occur.
4. Imbibition—Imbibition is the swelling of tissues, alive or dead, to several times their original volume. This is a result of the electrical charges on materials in suspension (colloidal) such as minerals, cellulose and starches attracting highly polar water molecules which then move into the cell. This swelling process is the initial step in the germination of seeds.
5. Active Transport—Active transport is the energy assisted movement of substances against a diffusion or electrical gradient. This process requires enzymes and a 'proton-pump' embedded in the plasma membrane. The pumps are energized by ATP molecules—a cellular energy storage molecule.

Water and its Movement Through the Plant

Roughly 90% of the water that enters a plant is lost via transpiration. Transpiration is the loss of water vapor through the leaves, just to refresh you. In addition, less than 5% of the water entering the plant is lost through the cuticle. Water is vital to plant life, not just for turgor pressure reasons, but much of the cellular activities occur in the presence of water molecules and the internal temperature of the plant is regulated by water. Recall that the xylem pathways go from the smallest part of the youngest roots all the way up the plant and out to the tip of the smallest and newest leaf. This internal plumbing system, paired with phloem and its nutrient transportation system, maintains the water needs and resources in the plant. The issue of the processes by which water is raised through columns—of considerable height at times—has been studied and debated for years in botany circles. The end result is the cohesion-tension theory.

The Cohesion-Tension Theory

Polar water molecules adhere to the walls of xylem tracheids and vessels and cohere to each other which allows an overall tension and form 'columns' of water in the plant.

The columns of water move from root to shoot and the water content of the soil supplies the 'columns' with water that enters the roots via osmosis. The difference between the water potentials of the soil and the air around the stomata are capable of producing enough force to transport water through the plant—from bottom to top and thus goes the cycle.

Transport of Food Substances (Organic Solutes) in Solution

Phloem is responsible for transporting food substance throughout the plant. As with water movement in plants, the movement of organic solutes in plants has been studied and debated for years. The currently accepted hypothesis is the pressure-flow hypothesis for the translocation of solutes.

The Pressure-Flow Hypothesis

This is essentially a source and sink hypothesis. Food substances that are in solution flow from a source, which is generally where water is taken up by osmosis (roots; food storage tissues, such as root cortex or rhizomes; and food producing tissues such as mesophyll in leaves), and the food substances are then given up at a destination or a sink where the food resources will be utilized in growth. The idea is that the organic solutes are moved along concentration gradients existing between sources and sinks.

At the source, phloem-loading occurs and sugars are moved by active transport into the sieve tubes of the smallest veins. The overall water potential in the sieve tube drops and then water enters the phloem cells via osmosis. The resulting turgor pressure from the movement of the water is enough to drive the solution through the sieve network to the sink . The sugar is unloaded at the sink via active transport and water then exits the ends of the sieve tubes. The pressure drops as the water exits, which causes a mass flow from the higher pressure at the source to the now lowered pressure at the sink. Much of the water that exits the sieve tubes will diffuse back into the xylem where it can be recirculated, transpired once it reaches the source. In a nutshell the mass flow is caused by drops in turgor pressure at the sink as the sugar molecules are removed. This generates the next push of materials toward the sink.

Regulation of Transpiration

It is the responsibility of the stomata to regulate transpiration and gas exchange via the actions of the guard cells. The pores of the stomata are closed when turgor pressure in the guard cells is low, and they are open when turgor pressure is high. Changes occur when light intensity, carbon dioxide concentration or water concentration change. The guard cells of the stomata use energy to take up potassium ions from adjacent epidermal cells.The uptake opens the stomata because water potential in the stomata drops and water moves into the guard cells and increases turgor pressure. When the potassium ions are released, the water then leaves the cells as the water potential shifts again. There is evidence that stomata will close with water stresses, but there also seems to be some indication that hormones are involved cause a loss of potassium ions from the guard cells and thus a pore closure.

Most plants keep their stomata open during the day and close them at night. However, there are plants that do the opposite and open their stomata during the night when overall water stress is lower. These plants have a specialized form of photosynthesis called CAM photosynthesis since the standard source of carbon dioxide is shut off as the stomata are closed during daylight hours. There are desert plants that are able to store carbon dioxide in their vacuoles in the form of organic acids that are converted back into carbon dioxide during the daytime for standard photosynthetic processes. As mentioned earlier, there are also adaptations such as sunken stomata which reduce the loss of water. Submerged or partially submerged plants generally do not have stomata on the underwater portions of their leaves.

High humidity will reduce transpiration rates while low humidity accelerates the process. There is a direct correlation between temperature and water movement out of the leaf. At high temperatures the rate of transpiration increases, while the opposite occurs at lower temperatures.

Mineral Requirements for Growth

Many external factors will affect growth rates and quality. The minerals available in the local soil is one such source of external input. Essential plant elements include: carbon, hydrogen, oxygen, phosphorus, potassium, nitrogen, sulphur, calcium, iron, magnesium sodium, chlorine, copper, manganese, cobalt, zinc, molybdenum, and boron to name the most common. Other minerals are required but they vary greatly from plant to plant. For example some algae need large amounts of iodine and silicon, while some loco weed species need selenium—which is poisonous to cattle on its own.

Macronutrients and Micronutrients

When any of these elements are lacking in the soil and the deficiencies are not compensated for by adding fertilizer compounds of compost the plant will demonstrate characteristic symptoms of mineral deficiencies. Most commercial fertilizers are some ratio of nitrogen, phosphorus and potassium and thus are able to compensate for a wide variety of issues. As an example of uses for the essential element in plants we will look at a few elements and how they are utilized:

Nitrogen—used in the building of proteins, nucleic acids and chlorophyll

Potassium—responsible in the process of enzyme activation, usually found in the Meristems

Calcium—vital part of the middle lamella and has a direct role in the movement of substances through cell membranes

Phosphorus—vital role in respiration and cellular division also used in the synthesis of energy compounds—ATP and ADP

Magnesium—central component of the chlorophyll molecule and involved in enzyme activation

Sulphur—structural component of many amino acids

Iron—integral in chlorophyll production and plays role in respiration

Manganese—enzyme activator

Boron—role in calcium ion use, not clearly understood

Plant Metabolism

Plants are responsible for incredible feats of molecular transformation. The processes are always being studied, but there are a few basic things that are well understood at this point in history. We will be looking at photosynthesis and respiration in some detail, as always, if you have additional questions please post them on the forum.

Photosynthesis

Photosynthesis is the process by which light energy is captured, converted and stored in simple sugar molecule. This process occurs in chloroplasts and other parts of green organisms. It is a backbone process, in the sense that all life on earth depends on it's functioning. The following equation sums up the process:

$$6CO_2 + 12\ H_2O + \text{light energy} \rightarrow C_6H_{12}O_6 + 6O_2 + 6H_2O$$

carbon water glucose oxygen water

Carbon Dioxide (CO_2)

The earth's atmosphere contains approximately 79% nitrogen, 20% oxygen and the remaining 1% is a mixture of less common gases—including 0.039% carbon dioxide. Carbon dioxide in the atmosphere reaches plant mesophyll via the stomata. The carbon dioxide dissolves on the thin film of water that covers the outside of cells. The carbon dioxide then diffuses through the cell wall into the cytoplasm in order to reach the chloroplasts. The oceans hold a large reservoir of carbon dioxide, which keeps the atmospheric levels essentially constant. Although there are some indicators that the atmospheric levels of CO_2 are rising and adding to the global warming issue. That is a whole other topic though.

Water

Water is plentiful on earth, however, it may or may not be plentiful at the location of each individual plant. Therefore, plants will close their stomata, if need be, which reduces the CO_2 supply to the mesophyll. Not even 1% of the water that is absorbed by plants is used in photosynthesis, the remainder is either transpired or incorporated into protoplasm, vacuoles or other cell materials. The water utilized in photosynthesis is the source of oxygen released as a photosynthetic byproduct.

Light

Light has a dual nature, in that it exhibits properties of both waves and particles. The energy from the sun comes to earth in various wavelengths, the longest being radio waves and the shortest are gamma rays. Approximately 40% of the radiant energy the earth receives from the sun is visible light. Visible light ranges from red, 780 nanometers to violet, 390 nanometers. The violet to blue and red to orange ranges are the most often used in photosynthesis. Most light in the green range is reflected. Of the visible light that reaches a leaf, approximately 80% is absorbed. Light intensity varies widely. Time of day, temperature, season of year, altitude, latitude and other atmospheric conditions all play roles in the intensity of the radiant energy that will reach the earth and it's organisms. High intensity light isn't necessarily a beneficial thing for plants. In high intensity light, photorespiration may occur, which is a type of respiration that uses oxygen and releases carbon dioxide but differs from standard aerobic respiration in the pathways that it utilizes.

Chlorophyll

A few things to know about chlorophyll before we get into the nitty gritty of photosynthesis and respiration. There are more than one type of chlorophyll, however, they all have one atom of magnesium in the centre. In some ways the chlorophyll is quite analogous to the heme structure in hemoglobin (the iron containing pigment that carries oxygen in blood). Chlorophyll has a long lipid tail that anchors the molecule in the lipid layers of the thylakoid membranes—recall that thylakoids are coin-like discs in stacks within the stroma of the chloroplasts. The chloroplasts of most plants contain two types of chlorophyll imbedded in the thylakoid membranes. The formula for bluish-green chlorophyll a is $C_{55}H_{72}MgN_4O_5$ and the formula for yellow-green chlorophyll b is $C_{55}H_{70}MgN_4O_6$. In general, most a chloroplast has about three times as much chlorophyll a than b. The main role of chlorophyll b is to broaden the spectrum of light available for photosynthesis: chlorophyll b absorbs light energy and transfers the energy to a chlorophyll a molecule. Other pigments are contained in chlorophyll c, d, and e and take the place for chlorophyll b in some cases. Note that all the chlorophyll molecules are related to each other and differ only slightly in molecular structure. Light-harvesting complexes contain 250 to 400 pigment molecules and are referred to as a photosynthetic unit. There are countless numbers of these units spread throughout the grana of a chloroplast. In the chloroplasts of green plants, two types of these harvesting units operate together in order to bring about the first phase of photosynthesis.

The photosynthetic process occurs in two successive processes: the light reactions and the carbon-fixing reactions.

The Light Reactions: The light reactions involve light striking the chlorophyll molecules embedded in the thylakoids of chloroplasts. The subsequent reaction results in the conversion of some light energy to chemical energy. In the light reactions, water molecules are split apart into hydrogen ions and electrons and oxygen gas is released. In addition, ATP (adenosine triphosphate) molecules are created and the hydrogen ions derived from the water molecules are involved in "loading" NADP which carries the hydrogen as NADPH. NADPH is integral in providing the hydrogen ions used in the second series of major photosynthetic reactions: the carbon-fixing reactions.

The Carbon-fixing Reactions: The carbon-fixing reactions used to be called the dark reactions because light does not play a direct role in their functioning. The reactions take place in series outside of the grana in the stroma of the chloroplast. These reactions only occur if the end products of the light reactions are available for use. Depending on the plant involved, the carbon-fixing reactions may develop or progress in different ways. The most common type of carbon-fixing reactions in plants is the process called the Calvin cycle. In the Calvin cycle, carbon dioxide from the atmosphere is combined with a 5-carbon sugar—RuBP, or ribulose bisphosphate. The combined molecules are converted via several steps into a 6-carbon sugar, such as glucose. The ATP and NADPH molecules from the light reactions provide the energy and resources for the reactions. Some of the sugars produced are further combined into polysaccharides (strings of simple sugars) or are stored as starch within the plant. There are other variations, including the 4-carbon pathway which is usually found in desert plants (C_4 plants).

Before getting into respiration let's take a closer look at what happens in both the light reactions and the carbon-fixing reactions.

Nitty-gritty of Light Reactions

Einstein called the discrete particles of light photons. Particles (photons) and waves are both currently accepted aspects of light. The quantum (energy) of photons is different depending on what kind of light they are in. Longer wavelength light has lower photon energies, while light with shorter wavelengths have higher photon energies. As mentioned earlier, every pigment colour has a different distinctive pattern of light absorption—called the pigment's absorption spectrum. The energy levels of some of the pigment's electrons are raised when the pigment absorbs light. If energy is emitted immediately upon absorption, the effect is called fluorescence. The red part of light does this characteristically, as demonstrated when chlorophyll is placed in light it will appear red. If the absorbed energy is emitted as light after a delay, then the effect is called phosphorescence. The energy may be converted to heat or stored, as in photosynthesis within chemical bonds.

Oxidation-reduction Reactions

OIL RIG, a cute little mnemonic device to remember that oxidation is loss and reduction is gain. Perhaps better put, oxidation results in the net loss of an electron or electrons, while reduction results in a net gain of an electron or electrons. The electrons come from compounds within the process or donated in from previous processes. These types of chemical reactions are found scattered throughout the processes within photosynthesis and respiration.

Photosystems

The two types of photosynthetic units in most chloroplasts are what constitute photosystem I and photosystem II.

1. Photosystem I contains photosynthetic units with 200 or more molecules of chlorophyll a, small amounts of chlorophyll b, protein saddled carotenoid pigment, and a pair of specialized reaction-centre molecules of chlorophyll called P_{700}. All pigments in a photosystem are capable of absorbing photons, however, only the reaction-centre molecules can really utilize the light energy. The other pigments aren't worthless in the system, as they act sort of like an antenna in gathering and passing light energy along to the reaction-centre. Iron-sulphur complexed proteins initially receive electrons from P_{700} and serve as primary electron acceptors for the unit.
2. Photosystem II contains chlorophyll a, protein saddled beta-carotene, a small amount of chlorophyll b and special pair of reaction-centre molecules of chlorophyll a otherwise called, P_{680}. The photosystem has a primary electron acceptor called pheophytin or Pheo.

For the record, the 680 and 700 in the names of the reaction-centre molecules stands for the peaks in the absorption spectra of light waves of 680 nm and 700 nm.

Photolysis

A photon of light strikes the photosystem II reaction-centre, the P_{680} molecule to be exact near the inner surface of a thylakoid membrane. The received light energy excites an electron (boosts it to a higher energy level) which is an unstable reaction and thus most of the energy is lost to heat. Up to four photons at a time can strike the P680 molecule, however, it can only accept one electron at a time. The molecule of pheophytin picks up the excited electron, which then crosses the thylakoid membrane and is passed along to another acceptor called plastoquinone or Pq near the outside surface of the thylakoid membrane. Protein Z extracts electrons from water and replaces the ones lost by the P680 molecule. Protein Z contains manganese which is required in order to split water molecules. Simultaneously, as two water molecules are split and molecule of oxygen and four protons are produced. This enzyme-mediated water splitting process is called photolysis.

Photophosphorylation

Pq, the acceptor molecule, releases the excited electron into the care of an electron transport system that is sort of like a downhill bucket brigade. The transport system moves electrons extracted from water temporarily to a high-energy storage molecule called nicotinamide adenine dinucleotide phosphate ($NADP^+$). $NADP^+$ is an electron acceptor for photosystem. The transport chain is essentially iron-containing pigments, cytochromes, a copper containing protein called plastocyanin and other electron transferring molecules. As electrons are passed through the chain and protons are being shuffled through a coupling factor, ATP molecules are assembled from ADP and phosphate in a process called photophosphorylation.

A similar series of events occurs in photosystem I. After a photon of light strikes a P_{700} molecule, the resulting excited electron is passed along to an iron-sulphur molecule Fe-S which in turn passes it to another acceptor molecule ferrodoxin, (Fd). The ferrodoxin molecule releases the electron to a carrier molecule called flavin adenine dinucleotide (FAD) and then eventually on to $NADP^+$. A reduction occurs and NADP+ becomes NADPH. Electrons from photosystem II and the activities of the electron transport system replace any electrons removed from the P_{700} molecule. Because the electrons move in one direction, the movement of electrons from water to photosystem II to photosystem I to $NADP^+$ are said to be part of noncyclic electron flow. Any ATP that is produced are designated noncyclic phosphorylation.

It should be noted that photosystem I can operate independently of photosystem II. When this occurs, the electrons boosted from P_{700} reaction-centre molecules (photosystem I) are passed through an intermediary acceptor molecule called P_{430} and then on to the electron transport chain. This is rather then to the ferrodoxin and $NADP^+$. After being passed through the electron transport chain, the electron is dumped back into the reaction-centre of photosystem I. This process demonstrates cyclic electron flow and any ATP generated by cyclic electron flow is termed cyclic phosphorylation. Note, that no water molecules are split and no NADPH or oxygen is produced.

Chemiosmosis

Earlier we mentioned in passing a coupling factor. The enzyme necessary for mediation of the splitting of water molecules is on the inside of the thylakoid membrane. As a result of this, a proton gradient forms across the membrane and the movement of these protons is thought to be a source of energy for generating ATP. The motion is thought to be similar to molecular movement during osmosis and has hence been termed chemiosmosis. As the protons move across the membrane, they are assisted in crossing by protein channels called ATPase or coupling factor. Because of the proton movement, ADP and phosphate combine which produces ATP.

Nitty-gritty of Carbon-Fixing Reactions

Both ATP and NADPH are important products of the light reactions and both of them play roles in the synthesis of carbohydrates from atmospheric carbon dioxide. Although the carbon-fixing reactions do not require daylight, they generally are conducted during daylight hours as there is some indication that some of the enzymes required for the processes in carbon-fixing may require some level of light. These reactions take place in the stroma of the chloroplast.

Three known mechanisms of converting carbon dioxide to sugar.

1. The Calvin Cycle or the 3-carbon pathway—The Calvin cycle is the most common of the three mechanisms and has four main results:
2. With the assistance of the enzyme rubisco (RuBP carboxylase), six molecules of atmospheric carbon dioxide combine with six molecules of ribulose 1, 5-bisphosphate (RuBP)
3. The result of the first step is six unstable 6-carbon complexes, which immediately split into two 3-carbon molecules of 3-phosphoglyceric acid or 3PGA. This is the first stable compound in photosynthesis.
4. NADPH and ATP from the light-reactions, supply the energy required to convert the 3PGA to 12 molecules of glyceraldehydes 3-phosphate (GA3P), which is a 3-carbon sugar phosphate complex.
5. Finally, of the 12 molecules formed; 10 are restructured into six 5-carbon molecules of RuBP—the sugar that the process started with.

The sugars produced can either add to an increase in the sugar content (carbohydrate content) of the plant or they can be used in pathways that lead to the production of lipids and amino acids.

1. 4-Carbon Pathway—C_4 plants: These plants use a 4-carbon molecule called oxaloacetic acid in place of the 3-carbon 3-

phosphoglyceric acid used in step two of the Calvin cycle. Oxaloacetic acid is produced from a 3-carbon compound PEP—phosphoenolpyruvate and carbon dioxide. This process is enzyme mediated and occurs in the mesophyll cells of the leaf. Some species will convert the resulting oxaloacetic acid to aspartic, malic or other acids.

Note that the acids do not substitute for 3PGA. The 4-carbon acids migrate to the bundle sheaths surrounding the vascular bundles, where they are further converted to pyruvic acid and carbon dioxide. In returning to the mesophyll cells and interacting with ATP molecules, the pyruvic acids molecules are able to produce additional PEP. In the bundle sheath cells, the carbon dioxide formed converts into 3PGA and other molecules, by combining with RuBP. The other molecules formed are similar to the other ones formed in the Calvin cycle. The C_4 cycle furnishes carbon dioxide to the Calvin cycle in a more roundabout way than the C_3 pathway, but there is an advantage to this extra pathway. The extra pathway greatly reduces photorespiration in C_4 plants, and this is a good thing because photorespiration is in direct competition with the Calvin cycle and even takes place in the light while the Calvin cycle is functioning. During photorespiration, RuBP reacts with oxygen to create carbon dioxide; in contrast, during photosynthesis RuBP and carbon dioxide are used to form carbohydrates. C_4 plants are able to pick up carbon dioxide in very low concentrations via the mesophyll cells. The Calvin cycle occurs in the bundle sheath where carbon dioxide is readily available. Because of the location, the enzyme rubisco will be in a prime spot to catalyze the reaction between RuBP and carbon dioxide, rather than oxygen. As a result C_4 plants have photosynthetic rates that are two to three times higher than C_3 plants. There are a few other characteristic features of C_4 plants worth noting:

1. C_4 plants have two types of chloroplasts and an alternate pathway for using carbon dioxide. C_3 plants only have one type of chloroplast and one pathway. Chloroplasts with starch grains and are large with very little grana, and sometimes none, in the bundle sheath cells. In the mesophyll, the small, but numerous chloroplasts have no starch grains and contain highly developed grana.
2. PEP carboxylase is found in high concentration in the mesophyll cells which permits ready conversion of carbon dioxide to carbohydrate at lower concentrations than does rubisco (in bundle sheath cells) of the Calvin cycle.

3. The temperature ranges for C_4 plants are much higher than C_3 plants which enables C_4 plants to live well in conditions that would likely kill a C_3 plant.
1. CAM Photosynthesis—Crassulacean acid metabolism is a modified photosynthetic system that is somewhat similar to C_4 photosynthesis in that 4-carbon compounds are produced during the carbon-fixing reactions. CAM plants accumulate malic acid in their chlorenchyma tissues at night, which is converted back to carbon dioxide during the day. In the daytime, malic acid diffuses out of the vacuoles and is converted to carbon dioxide for use in the Calvin cycle. PEP carboxylase is responsible for converting the carbon dioxide plus PEP to malic acid at night. This modification allows for a greater amount of carbon dioxide to be converted to carbohydrate during the day than would be otherwise converted given the conditions CAM plants generally grow in. CAM plants generally close their stomata during the day in order to reduce water loss. There are more than 20 families that contain CAM plants, including cacti, stonecrops, orchids, bromeliads and many succulents growing in regions of high light intensity. There are some succulents that do not have CAM photosynthetic capabilities, as well as non-succulents that do have the ability.

There are great resources available that go into even greater detail on these reactions. If you are interested in these titles, please don't be afraid to ask on the forum for direction.

Respiration

Respiration is the group of processes that utilizes the energy that is stored through the photosynthetic processes. The steps in respiration are small enzyme-mediated steps tha release tiny amounts of immediately available energy, the energy released is usually stored in ATP molecules which allow for even more efficient use of an organism's energy. Respiration occurs in the mitochondria and cytoplasm of cells.

There are several forms of respiration: aerobic—which requires oxygen, anaerobic—which occurs in the absence of oxygen, and fermentation—which also occurs in the absence of oxygen.

Aerobic respiration is the most common form of respiration and cannot be completed without oxygen gas. The controlled release of energy is the main event in aerobic respiration.

Certain types of bacteria and other organisms without oxygen gas carry on anaerobic respiration and fermentation. Compared to aerobic respiration the amount of energy released is quite small. The main difference between aerobic respiration and fermentation is in the way hydrogen is released and combined with other substances. Two very common forms of fermentation are summed up by the following equations:

$C_6H_{12}O_6$ -> (with enzymes)-> $2C_2H_5OH$ + $2CO_2$ + energy (ATP)
glucose ethyl alcohol carbon dioxide

$C_6H_{12}O_6$ -> (with enzymes) -> $2C_3H_6O_3$ + energy (ATP)
glucose lactic acid

Note the first equation is particularly valuable to the brewing industry.

Major Steps in Respiration: Glycolysis—the first step does not require oxygen gas (O_2) and takes place in the cytoplasm. The glycolytic phase is subdivided into three main steps and several smaller ones. Each step is mediated by an enzyme. A small amount of energy is released and hydrogen atoms are removed from compounds derived from glucose. The main gist of the steps are:

A. the glucose molecules goes through several steps and becomes a double phosphorylated fructose molecule.

B. The 6-carbon fructose molecule is split into two 3-carbon fragments, each with a phosphate, GA3P

C. Hydorgen, energy and water are removed from the GA3P molecules leaving pyruvic acid.

Glycolysis requires two molecules of ATP to get the process started. In the processes, four ATP molecules are created, with a net gain of 2 ATP molecules at the end of glycolysis. The hydrogen ions and electrons that are released are held by an acceptor molecule called NAD—nicotinamide adenine dinucleotide. The overall end products of gylcolysis is: 2-ATP molecules, 2-NADH molecules, and pyruvic acid.

The next step depends on the kind of respiration involved—aerobic, true anaerobic or fermentation.

Aerobic Respiration (with oxygen present)

1. The Krebs Cycle (or citric acid cycle)—The Krebs cycle takes place in the fluid matrix of the cristae compartments of the mitochondria. It is called the citric acid cycle because of all the

intermediate acids in the cycle. The pyruvic acid product of glycolysis is restructured, some of the CO_2 is lost and becomes acetyl CoA which then dumps into the Krebs cycle. During the restructuring of pyruvic acid, a molecule of NADH is produced. The Krebs cycle removes energy, CO_2 and hydrogen from acetyl CoA via enzyme mediated reactions of organic acids.

The hydrogen removed during the Krebs cycle is picked up by FAD and NAD acceptor molecules. The end result of the metabolizing of two acetyl CoA molecules in the Krebs cycle is: 2-ATP molecule, oxaloacetic acid (to further drive the cycle), 6-$NADH_2$ molecules, 2-$FADH_2$ molecules and $2CO_2$ molecules.

The NAD and FAD molecules and the hydrogens that they carry will be dumped into the next step in respiration in order to extract the energy from the molecules.

1. The Electron Transport Chain—The electron transport chain is a bit like a bucket brigade in that the chain passes the molecules along until the job is done. Energy is released as the hydrogen and electrons from the NAD^+ and FAD^+ carrier molecules is dumped into the system. When the electrons reach the end of the chain they pick up an oxygen and water is released. ATP is produced by oxidative phosphorylation during the action of the electron transport chain. This occurs essentially like chemiosmosis.

As a whole, from glycolysis to finish aerobic respiration yields the following:

Glycolysis:

4-molecules of ATP

+2-molecules of NADH—which yields 4-ATP in the ETS

8-molecules of ATP net

-2 ATP molecules to start the glycolysis process

6- ATP molecules

Conversion of pyruvic acid to acetyl CoA:

2-molecules of NADH—yields 6 ATP in the ETS

Krebs Cycle:

2-molecules of ATP

2-molecules of $FADH_2$—which yields 4-ATP in the ETS
6-molecules of $NADH_2$—which yields 18-ATP in the ETS

———————

Total ATP yield: 36

The 36 resulting ATP molecules represent approximately 39% of the energy in a molecule of glucose. Compared to each other, aerobic respiration is about six times as efficient as anaerobic respiration.

Anaerobic respiration and fermentation result in a net gain of 2-ATP molecules from glycolysis. It should be noted, that the by-products of these processes, lactic acid and alcohol, will eventually kill the organism if the products are not digested.

Factors Regulating Rate of Respiration

Temperature—To a point, the higher the temperature the faster respiration occurs. At some temperature, enzymes will become inactivated, although there are thermophilic (heat-loving) organisms that do quite well in high-temperature environments. Energy from sugar is released faster as the rate of respiration increases which results in a net weight loss. Plants offset the weight loss by increasing photosynthetic production of sugar. Note that during respiration, some of the energy is lost as heat, which results in an overall increase in organism temperature—not necessarily detectible by human hands.

Water—Enzymes generally operate in the presence of water, and reduced water in a plant will reduce the rate of respiration. Seeds usually have a water content of less than 10%, while mature living cells usually are in excess of 90% water. Seeds keep better if they are kept dry as the respiration rate remains quite low. However, if a seed comes into contact with water and via imbibition swells, the respiration rate will skyrocket. The temperature could increase to the point of killing the seeds. Spontaneous combustion can occur from the respiration generated heat when a fungi or bacterium is permitted to grow on wet seeds. Kind of a neat little trivia fact to tuck away.

Oxygen—Oxygen is an important regulator of respiration. If oxygen is drastically reduced, respiration may drop off to the point of retarding growth or death. Low oxygen concentrations can lead to the onset of fermentation processes.

Assimilation and Digestion

Assimilation is the conversion of the sugar produced by photosynthesis to fats, proteins, complex carbohydrates and other substances. While digestion is the breakdown of large insoluble molecules by hydrolysis to smaller soluble forms that can be transported to various parts of the plant.

Summary of key differences between photosynthesis and respiration:

- Photosynthesis
- Energy stored in sugar molecules
- Carbon dioxide and water used
- Increases weight
- Requires light
- Occurs in chlorophyll
- In green organisms, produces oxygen
- With light energy, produces ATP
- Respiration
- Energy released from sugar molecules
- Carbon dioxide and water released
- Decreases weight
- Can occur in light or darkness
- Occurs in all living cells
- Uses oxygen (aerobic respiration)
- With energy released from sugar, produces ATP.

Growth and Plant Hormones

Growth

All living organisms begin in the same form: as a single cell. That cell will divide and the resulting cells will continue dividing and differentiate into cells with various roles to carry out within the organism. This is life and plants are no different. Plant growth can be determinate or indeterminate, meaning some plants will have a cycle of growth then a cessation of growth, breakdown of tissues and then death (think of a radish plant or a tomato plant) while others (think of a giant cedar tree) will grow and remain active for hundreds of years. A tomato plant is fairly predictable and is said to have determinate growth, while the cedar tree has indeterminate growing potential. Development refers to the growth and differentiation of cells into tissues, organs and organ systems. This again all begins with a single cell.

Plant Growth Regulators and Enzymes

Genetic information directs the synthesis and development of enzymes which are critical in all metabolic process within the plant. Most enzymes are proteins in some form or another, are produced in very minute quantities and are produced on site—meaning they are not transported from one part of the organism to another. Genetic information also regulates the production of hormones, which will be addressed shortly. The major difference is that hormones are transported from one part of the plant to another as needed. Vitamins vital in the activation of enzymes and are produced in the cytoplasm and membranes of plant cells. Animals and humans utilize plants in order to provide some vitamin resources. In general, hormone and vitamin effects are similar and are difficult to distinguish in plants, and both are referred to in general as plant growth regulators.

Plant Hormones

The growth and development of a plant are influenced by genetic factors, external environmental factors, and chemical hormones inside the plant. Plants respond to many environmental factors such as light, gravity, water, inorganic nutrients, and temperature.

Groups of Hormones

Plant hormones are chemical messengers that affect a plant's ability to respond to its environment. Hormones are organic compounds that are effective at very low concentration; they are usually synthesized in one part of the plant and are transported to another location. They interact with specific target tissues to cause physiological responses, such as growth or fruit ripening. Each response is often the result of two or more hormones acting together.

Because hormones stimulate or inhibit plant growth, many botanists also refer to them as plant growth regulators. Many hormones can be synthesized in the laboratory, increasing the quantity of hormones available for commercial applications. Botanists recognize five major groups of hormones: auxins, gibberellins, ethylene, cytokinins, and abscisic acid.

Auxins

Auxins are hormones involved in plant-cell elongation, apical dominance, and rooting. A well known natural auxin is indoleacetic acid, or IAA which is produced in the apical meristem of the shoot. Developing seeds produce IAA, which stimulates the development of a fleshy fruit. For example, the removal of seeds from a strawberry prevents the fruit from enlarging. The application of IAA after removing the seeds causes the fruit to enlarge normally. IAA is produced in actively growing shoot tips and developing fruit, and it is involved in elongation. Before a cell can elongate, the cell wall must become less rigid so that it can expand. IAA triggers an increase in the plasticity, or stretchability, of cell walls, allowing elongation to occur.

Synthetic Auxins

Chemists have synthesized several inexpensive compounds similar in structure to IAA. Synthetic auxins, like naphthalene acetic acid, of NAA, are used extensively to promote root formation on stem and leaf cuttings. Gardeners often spray auxins on tomato plants to increase the number of fruits on each plant. When NAA is sprayed on young fruits of apple and olive trees, some of the fruits drop off so that the remaining fruits grow larger.

When NAA is sprayed directly on maturing fruits, such as apples, pears and citrus fruits, several weeks before they are ready to be picked; NAA prevents the fruits from dropping off the trees before they are mature. The fact that auxins can have opposite effects, causing fruit to drop or preventing fruit from dropping, illustrates an important point. The effects of a hormone on a plant often depend on the stage of the plant's development.

NAA is used to prevent the undesirable sprouting of stems from the base of ornamental trees. As previously discussed, stems contain a lateral bud at the base of each leaf. IN many stems, these buds fail to sprout as long as the plant's shoot tip is still intact. The inhibition of lateral buds by the presence of the shoot tip is called apical dominance. If the shoot tip of a plant is removed, the lateral buds begin to grow. If IAA or NAA is applied to the cut tip of the stem, the lateral buds remain dormant. This adaptation is manipulated to cultivate beautiful ornamental trees. NAA is used commercially to prevent buds from sprouting on potatoes during storage.

Another important synthetic auxin is 2,4-D, which is an herbicide, or weed killer. It selectively kills dicots, such as dandelions and pigweed, without injuring monocots, such as lawn grasses and cereal crops. Given our major dependence on cereals for food; 2,4-D has been of great value to agriculture. A mixture of 2, 4-D and another auxin, called Agent Orange, was used to destroy foliage in the jungles of Vietnam. A non-auxin contaminant in Agent Orange has caused severe health problems in many people who were exposed to it.

Gibberellins

In the 1920's scientists in Japan discovered that a substance produced by the fungus *Gibberella* caused fungus-infected plants to grow abnormally tall. The substance, named gibberellin, was later found to be produced in small quantities by plants themselves. It has many effects on a plant, but primarily stimulates elongation growth. Spraying a plant with gibberellins will usually cause the plant to grow to a larger than expected height, i.e. greater than normal.

Like auxins, gibberellins are a class of hormones that have important commercial applications. Almost all seedless grapes are sprayed with gibberellins to increase the size of the fruit and the distance between fruits on the stems. Beer makers use gibberellins to increase the alcohol content of beer by increasing the amount of sugar produced in the malting process. Gibberellins are also used to treat seeds of some food crops because they will break seed dormancy and promote uniform germination.

Ethylene

The hormone ethylene is responsible for the ripening of fruits. Unlike the other four classes of plant hormones, ethylene is a gas at room temperature. Ethylene gas diffuses easily through the air from one plant to another. The saying "One bad apple spoils the barrel" has its basis in the effects of ethylene gas. One rotting apple will produce ethylene gas, which stimulates nearby apples to ripen and eventually spoil because of over ripening.

Ethylene is usually applied in a solution of ethephon, a synthetic chemical that breaks down and releases ethylene gas. It is used to ripen bananas, honeydew melons and tomatoes. Oranges, lemons, and grapefruits often remain green when they are ripe. Although the fruit tastes good, consumers often will not buy them, because oranges are supposed to be orange, right? The application of ethylene to green citrus fruit causes the development of desirable citrus colours, such as orange and yellow. In some plant species, ethylene promotes abscission, which is the detachment of leaves, flowers, or fruits from a plant. Cherries and walnuts are harvested with mechanical tree shakers. Ethylene treatment increases the number of fruits that fall to the ground when the trees are shaken. Leaf abscission is also an adaptive advantage for the plant. Dead, damaged or infected leaves drop to the ground rather than shading healthy leaves or spreading disease. The plant can minimize water loss in the winter, when the water in the plant is often frozen.

Cytokinins

Cytokinins promote cell division in plants. Produced in the developing shoots, roots, fruits and seeds of a plant, cytokinins are very important in the culturing of plant tissues in the laboratory. A high ratio of auxins to cytokinins in a tissue-culture medium stimulates root formation. A low ratio promotes shoot formation. Cytokinins are also used to promote lateral bud growth in flowering plants.

Abscisic Acid

Abscisic acid, or ABA, generally inhibits other hormones, such as the auxin IAA. It was originally thought to promote abscission, hence its name. Botanists now know that ethylene in the main abscission hormone. ABA helps to bring about dormancy in a plant's buds and maintains dormancy in its seeds. ABA causes the closure of a plant's stomata in response to drought. Water stressed leaves produce large amounts of ABA, which triggers potassium ions to be transported out of the guard cells. This causes stomata to close, and water is held in the leaf. It is too costly to synthesize ABA for commercial agriculture use.

Other Growth Regulators

Many growth regulators are widely used on ornamental plants. These substances do not fit into any of the five classes of hormones. For example, utility companies all over the country often apply growth retardants, chemicals that prevent plant growth, to trees in order to prevent them from interfering with overhead utility lines. If is less expensive to apply these chemicals than to prune the trees, not to mention safer for the utility workers. Also, azalea growers sometimes apply a chemical to the terminal buds rather than hand-pruning them. Scientists are still searching for a hormone to slow the growth of lawn grass so that it doesn't have to be mowed so often.

Plant Movements

Plants appear immobile because they are usually rooted in one place. However, time lapse photography reveals that parts of plants frequently move. Most plants move too slowly for the passerby to notice. Plants move in response to several environmental stimuli such as: light, gravity and mechanical disturbances. These movements fall into two groups: tropisms and nastic movements.

Tropisms

A tropism is a plant movement that is determined by the direction of an environmental stimulus. Movement toward an environmental stimulus is called a positive tropism, and movement away from a stimulus is called a negative tropism. Each kind of tropism is named for its stimulus. For example, a plant movement in response to light coming from one particular direction is called a phototropism. The shoot tips of a plant that grow toward the light source are positively phototropic.

Phototropism

Phototropism, as mentioned, is illustrated by the movement of sprouts in relation to light source direction. Light causes the hormone auxin to move tot he shaded side of the shoot. The auxin causes the cells on the shaded side to elongate more than the cells on the illuminated side. As a result, the shoot bends toward the light and exhibits positive phototropism. In some plant stems, phototropism is not caused by auxin presence or movement. In these instances, light causes the production of a growth inhibitor on the illuminated side of the shoot. Negative phototropism is sometimes seen in vines that climb on flat walls where coiling tendrils have nothing to coil around. These vines have stem tips that grow away from the light, or better put, toward the wall. This brings adventitious roots or adhesive discs in contact with the wall on which they can cling and climb.

Solar tracking is the motion of leaves or flowers as the follow the suns' movement across the sky. By continuously facing toward a light source, moving or not, the plant maximizes the light available for photosynthesis.

Thigmotropism

Thigmotropism is a plant growth response to touching a solid object. Tendrils and stems of vines, such as morning glories, coil when they touch an object. Thigmotropism allows some vines to climb other plants or objects, thus increasing its chance of intercepting light for photosynthesis. It is thought that an auxin and ethylene are involved in this response.

Gravitropism

Gravitropism is a plant growth response to gravity. A root usually grows downward and a stem usually grows upward; that is, roots are positively gravitropic and stems are negatively gravitropic. Like phototropism, gravitropism appears to be regulated by auxins. One hypothesis proposes that when a seedling is placed horizontally, auxins accumulate along the lower sides of the root and the stem. This concentration of auxins stimulates cell elongation along the lower side of the stem, and the stem grows upward. A similar concentration of auxins inhibits cell elongation in the lower side of the root, and thus the root grows downward.

Chemotropism

Chemotropism is a plant growth response to a chemical. After a flower is pollinated, a pollen tube grows down through the stigma and style and enters the ovule through the micropyle. The growth of the pollen tube in response to chemicals produced by the ovule is an excellent example of chemotropism.

Nastic Movements

Plant movements that occur in response to environmental stimuli, but that are independent of the direction of the stimuli are called nastic movements. These movements are regulated by changes in water pressure in certain plant cells.

Thigmonastic Movements

Thigmonastic movements are a type of nastic movements that occur in response to touching or shaking a plant. Many thigmonasties involve rapid plant movements, such as the closing of the leaf trap of a Venus flytrap plant or the folding of a plant's leaves in response to being touched. Some leaves of sensitive plants will fold within a few seconds after being touched. This movement is caused by the rapid loss of turgor pressure (water pressure) in certain cells, a process similar to that which occurs in guard cells in order to close stomata. Physical stimulation of the plant leaf causes potassium ions to be pumped out of the cells at the base of the leaflets and petioles. Water then moves out of the cells by osmosis. As the cells shrink, the plant leaves move. It is believed that the folding of a plant's leaves in response to touch is to discourage insect feeding.

In addition, thigmonastic movements help prevent water loss in plants. When the wind blows across a plant, the rate of transpiration is increased. If the leaves of a plant fold in response to the "touch" of the wind, water loss is reduced.

Nyctinastic Movements

Nyctinastic movements are plant movements in response to the daily cycle of light and dark. Nyctinastic movements involve the same type of osmotic mechanism as thigmonastic movements, but the changes in turgor pressure are more gradual. Nyctinastic movements occur in many plants. Examples of plants that demonstrate these movements include honeylocust trees, silk trees and bean plants. The prayer plant gets its name from the fact that its leaf blades are vertical at night, resembling praying hands. During the day, however, the leaf blades are positioned horizontally. Carolus Linnaeus planted a "flower clock" made of different species of plants with nyctinastic movements. The movements of each plant species occurred at a specific time of day when the light was right for the plant.

Seasonal Responses

In nontropical areas, plant responses are strongly influenced by seasonal changes. For example, many trees shed their leaves in the fall, and most plants flower only at certain times of the year. Plants are able to sense seasonal changes. Although temperature changes are involved in some case and to certain degrees, plants mark the seasons primarily by sensing changes in night length.

Photoperiodism

A plant's response to changes in the length of days and nights is called photoperiodism. Photoperiodism affects many plant processes, including the formation of storage organs and bud dormancy. However, the most studied photoperiodic process is flowering. Some plants require a particular night length to flower. In other species, a particular night length merely makes a plant flower sooner than it otherwise would.

Critical Night Length

It has been discovered that the important factor in flowering is the amount of darkness, or night length, that a plant receives. Each plant species has its own specific requirements for darkness, called the critical night length. Although it is now understood that night length, and not day length, regulates flowering, the terms short-day plant and long-day plant are still used. A short-day plant flowers when the days are short and the nights are long. Conversely, a long-day plant flowers when the days are long and the nights are short compared to the requirements of another plant.

Responding to Day Length and Night Length

Plants can be divided into three groups, depending on their response tot he photoperiod, which again acts a season indicator.

One group, called day-neutral plants (DNPs) are not affected by day length. Examples of DNPs for flowering include tomatoes, dandelions, roses, corn, cotton and beans.

Short-day plants (SDPs) flower in the spring of fall, when the day length is short. For example ragweed flowers when the days are shorter than 14 hours and poinsettias flower when the days are shorter than 12 hours. Chrysanthemums, goldenrods, and soybeans are SDPs for flowering.

Long-day plants (LDPs) flower when the days are long, usually in summer. For example, wheat flowers only when the days are longer than 10 hours. Radishes, asters, petunias, and beets are LDPs for flowering.

Phytochrome Regulation in Plants

Plants monitor changes in day length with a bluish, light-sensitive protein pigment called phytochrome. Phytochrome exists in two forms, based on the wavelength of the light that it absorbs. It is generally produced in meristematic tissues in very minute amounts. The two stable forms can be converted to each other by absorbing light. P_{red} (P_r) which absorbs red light and $P_{far\text{-}red}$ (P_{fr}) which absorbs far-red light. In the daylight more P_r is converted to P_{fr} (the active form) than vice versa. P_{fr} will convert back to P_r over several hours in the dark where it would be stable indefinitely. The conversion in light is almost instantaneous. The phytochrome mechanism is what transforms the crook in the hypocotyls of the emerging seedling into a straight stalk. Stem elongation appears to be inhibited by $P_{fr.}$ However, if light levels are low, the shaded stems of a tree for example, more far-red light will reach them and cause the conversion to P_r which lowers inhibition and allows the stems to grow longer and out from under the shade.

Vernalization

Vernalization is the low-temperature stimulation of flowering. Vernalization is important for fall-sown grain crops, such as winter wheat, barley and rye. For example, wheat seeds are sown in the fall and survive the winter as small seedlings. Exposure to cold weather causes the plants to flower in the early spring, and an early crop is produced. If the same wheat is sown in the spring, it will take about two months longer to produce a crop. Thus, cold temperatures are not absolutely required for most crops, but they do expedite flowering. Farmers often use vernalization to grow and harvest their crops before a summer drought sets in and stunts growth.

A biennial plant is a plant that lives for two years, usually producing flowers and seeds during the second year. Biennial plants, such as carrots, beets, celery and foxglove, survive their first winter as short plants. In the spring their flowering stem elongates rapidly, a process called bolting. Most biennials must receive cold weather to vernalize before they flower during the second year. They will then die after flowering. Treating a biennial with gibberellin is sometimes a substitute for cold temperatures in vernalization, and will stimulate the plant to grow.

Fall Colours

Some tree leaves are noted for their spectacular fall colour display. The changing fall colours are caused primarily by a photoperiodic response but also by a temperature response. As nights become longer in the fall, leaves stop producing chlorophyll. As the chlorophyll chemically degrades, it is not replaced. Other leaf pigments, the carotenoids, become visible and the green/orange splotches become more visible as the green chlorophyll turns orange. Carotenoids include the orange carotenes and the yellow xanthophylls. Anthocyanins produce the deep red and purplish-red colours in the fall display.

3

Cell Membrane

The *'cell membrane'* (also known as the plasma membrane or cytoplasmic membrane) is a biological membrane that separates the interior of all cells from the outside environment. The cell membrane is selectively permeable to ions and organic molecules and controls the movement of substances in and out of cells. The basic function of the cell membrane is to protect the cell from its surroundings. It consists of the phospholipid bilayer with embedded proteins. Cell membranes are involved in a variety of cellular processes such as cell adhesion, ion conductivity and cell signalling and serve as the attachment surface for several extracellular structures, including the cell wall, glycocalyx, and intracellular cytoskeleton. Cell membranes can be artificially reassembled.

Function

The cell membrane or plasma membrane surrounds the cytoplasm of living cells, physically separating the intracellular components from the extracellular environment. Fungi, bacteria and plants also have the cell wall which provides a mechanical support for the cell and precludes the passage of larger molecules. The cell membrane also plays a role in anchoring the cytoskeleton provide shape to the cell, and in attaching to the extracellular matrix and other cells to help group cells together to form tissues.

The membrane is selectively permeable and able to regulate what enters and exits the cell, thus facilitating the transport of materials needed for survival. The movement of substances across the membrane can be either "passive", occurring without the input of cellular energy, or active, requiring the cell to expend energy in transporting it. The membrane also maintains the cell potential. The cell membrane thus works as a selective filter that allows only certain things to come inside or go outside the cell. The cell employs a number of transport mechanisms that involve biological membranes:

1. Passive osmosis and diffusion: Some substances (small molecules, ions) such as carbon dioxide (CO_2) and oxygen (O_2), can move across the plasma membrane by diffusion, which is a passive transport process. Because the membrane acts as a barrier for certain molecules and ions, they can occur in different concentrations on the two sides of the membrane. Such a concentration gradient across a semipermeable membrane sets up an osmotic flow for the water.
2. Transmembrane protein channels and transporters: Nutrients, such as sugars or amino acids, must enter the cell, and certain products of metabolism must leave the cell. Such molecules are pumped across the membrane by transmembrane transporters or diffuse through protein channels such as Aquaporins (in the case of water (H_2O)). These proteins, also called permeases, are usually quite specific, recognizing and transporting only a limited food group of chemical substances, often even only a single substance.
3. Endocytosis: Endocytosis is the process in which cells absorb molecules by engulfing them. The plasma membrane creates a small deformation inward, called an invagination, in which the substance to be transported is captured. The deformation then pinches off from the membrane on the inside of the cell, creating a vesicle containing the captured substance. Endocytosis is a pathway for internalizing solid particles (cell eating or phagocytosis), small molecules and ions (cell drinking or pinocytosis), and macromolecules. Endocytosis requires energy and is thus a form of active transport.
4. Exocytosis: Just as material can be brought into the cell by invagination and formation of a vesicle, the membrane of a vesicle can be fused with the plasma membrane, extruding its contents to the surrounding medium. This is the process of exocytosis. Exocytosis occurs in various cells to remove undigested residues of substances brought in by endocytosis, to secrete substances such as hormones and enzymes, and to transport a substance completely across a cellular barrier. In the process of exocytosis, the undigested waste-containing food

vacuole or the secretory vesicle budded from Golgi apparatus, is first moved by cytoskeleton from the interior of the cell to the surface. The vesicle membrane comes in contact with the plasma membrane. The lipid molecules of the two bilayers rearrange themselves and the two membranes are, thus, fused. A passage is formed in the fused membrane and the vesicles discharges its contents outside the cell.

Prokaryotes

Gram-negative bacteria have both a plasma membrane and an outer membrane separated by a periplasmic space. Other prokaryotes have only a plasma membrane. Prokaryotic cells are also surrounded by a cell wall composed of peptidoglycan (amino acids and sugars). Some eukaryotic cells also have cells walls, but none that are made of peptidoglycan.

Structures

Fluid Mosaic Model

According to the fluid mosaic model of S.J.Singer and G.L. Nicolson (1972), which replaced the earlier model of Davson and Danielli, biological membranes can be considered as a two-dimensional liquid in which lipid and protein molecules diffuse more or less easily. Although the lipid bilayers that form the basis of the membranes do indeed form two-dimensional liquids by themselves, the plasma membrane also contains a large quantity of proteins, which provide more structure. Examples of such structures are protein-protein complexes, pickets and fences formed by the actin-based cytoskeleton, and potentially lipid rafts.

Lipid Bilayer

Lipid bilayers form through the process of self-assembly. The cell membrane consists primarily of a thin layer of amphipathic phospholipids which spontaneously arrange so that the hydrophobic "tail" regions are isolated from the surrounding polar fluid, causing the more hydrophilic "head" regions to associate with the intracellular (cytosolic) and extracellular faces of the resulting bilayer. This forms a continuous, spherical lipid bilayer. Forces such as van der Waals, electrostatic, hydrogen bonds, and noncovalent interactions all contribute to the formation of the lipid bilayer. Overall, hydrophobic interactions are the major driving force in the formation of lipid bilayers.

Lipid bilayers are generally impermeable to ions and polar molecules. The arrangement of hydrophilic heads and hydrophobic tails of the lipid bilayer prevent polar solutes (ex. amino acids, nucleic acids, carbohydrates, proteins, and ions) from diffusing across the membrane, but generally allows for the passive diffusion of hydrophobic molecules. This affords the cell the ability to control the movement of these substances via transmembrane protein complexes such as pores, channels and gates.

Flippases and scramblases concentrate phosphatidyl serine, which carries a negative charge, on the inner membrane. Along with NANA, this creates an extra barrier to charged moieties moving through the membrane. Membranes serve diverse functions in eukaryotic and prokaryotic cells. One important role is to regulate the movement of materials into and out of cells. The phospholipid bilayer structure (fluid mosaic model) with specific membrane proteins accounts for the selective permeability of the membrane and passive and active transport mechanisms. In addition, membranes in prokaryotes and in the mitochondria and chloroplasts of eukaryotes facilitate the synthesis of ATP through chemiosmosis.

Membrane Polarity

The apical membrane of a polarized cell is the surface of the plasma membrane that faces inward to the lumen. This is particularly evident in epithelial and endothelial cells, but also describes other polarized cells, such as neurons. The basolateral membrane of a polarized cell is the surface of the plasma membrane that forms its basal and lateral surfaces. It faces outwards, towards the interstitium, and away from the lumen. Basolateral membrane is a compound phrase referring to the terms "basal (base) membrane" and "lateral (side) membrane", which, especially in epithelial cells, are identical in composition and activity. Proteins (such as ion channels and pumps) are free to move from the basal to the lateral surface of the cell or vice versa in accordance with the fluid mosaic model. Tight junctions join epithelial cells near their apical surface to prevent the migration of proteins from the basolateral membrane to the apical membrane. The basal and lateral surfaces thus remain roughly equivalent to one another, yet distinct from the apical surface.

Membrane Structures

Cell membrane can form different types of "supramembrane" structures such as caveola, postsynaptic density, podosome, invadopodium, focal adhesion, and different types of cell junctions.

These structures are usually responsible for cell adhesion, communication, endocytosis and exocytosis. They can be visualized by electron microscopy or fluorescence microscopy. They are composed of specific proteins, such as integrins and cadherins.

Cytoskeleton

The cytoskeleton is found underlying the cell membrane in the cytoplasm and provides a scaffolding for membrane proteins to anchor to, as well as forming organelles that extend from the cell. Indeed, cytoskeletal elements interact extensively and intimately with the cell membrane. Anchoring proteins restricts them to a particular cell surface — for example, the apical surface of epithelial cells that line the vertebrate gut — and limits how far they may diffuse within the bilayer.

The cytoskeleton is able to form appendage-like organelles, such as cilia, which are microtubule-based extensions covered by the cell membrane, and filopodia, which are actin-based extensions. These extensions are ensheathed in membrane and project from the surface of the cell in order to sense the external environment and/or make contact with the substrate or other cells. The apical surfaces of epithelial cells are dense with actin-based finger-like projections known as microvilli, which increase cell surface area and thereby increase the absorption rate of nutrients. Localized decoupling of the cytoskeleton and cell membrane results in formation of a bleb.

Composition

Cell membranes contain a variety of biological molecules, notably lipids and proteins. Material is incorporated into the membrane, or deleted from it, by a variety of mechanisms:

- Fusion of intracellular vesicles with the membrane (exocytosis) not only excretes the contents of the vesicle but also incorporates the vesicle membrane's components into the cell membrane. The membrane may form blebs around extracellular material that pinch off to become vesicles (endocytosis).
- If a membrane is continuous with a tubular structure made of membrane material, then material from the tube can be drawn into the membrane continuously.
- Although the concentration of membrane components in the aqueous phase is low (stable membrane components have low solubility in water), there is an exchange of molecules between the lipid and aqueous phases.

Lipids

The cell membrane consists of three classes of amphipathic lipids: phospholipids, glycolipids, and sterols. The amount of each depends upon the type of cell, but in the majority of cases phospholipids are the most abundant. In RBC studies, 30% of the plasma membrane is lipid.

The fatty chains in phospholipids and glycolipids usually contain an even number of carbon atoms, typically between 16 and 20. The 16- and 18-carbon fatty acids are the most common. Fatty acids may be saturated or unsaturated, with the configuration of the double bonds nearly always "cis". The length and the degree of unsaturation of fatty acid chains have a profound effect on membrane fluidity as unsaturated lipids create a kink, preventing the fatty acids from packing together as tightly, thus decreasing the melting temperature (increasing the fluidity) of the membrane. The ability of some organisms to regulate the fluidity of their cell membranes by altering lipid composition is called homeoviscous adaptation.

The entire membrane is held together via non-covalent interaction of hydrophobic tails, however the structure is quite fluid and not fixed rigidly in place. Under physiological conditions phospholipid molecules in the cell membrane are in the liquid crystalline state. It means the lipid molecules are free to diffuse and exhibit rapid lateral diffusion along the layer in which they are present. However, the exchange of phospholipid molecules between intracellular and extracellular leaflets of the bilayer is a very slow process. Lipid rafts and caveolae are examples of cholesterol-enriched microdomains in the cell membrane.

In animal cells cholesterol is normally found dispersed in varying degrees throughout cell membranes, in the irregular spaces between the hydrophobic tails of the membrane lipids, where it confers a stiffening and strengthening effect on the membrane.

Phospholipids Forming Lipid Vesicles

Lipid vesicles or liposomes are circular pockets that are enclosed by a lipid bilayer. These structures are used in laboratories to study the effects of chemicals in cells by delivering these chemicals directly to the cell, as well as getting more insight into cell membrane permeability. Lipid vesicles and liposomes are formed by first suspending a lipid in an aqueous solution then agitating the mixture through sonication, resulting in a vesicle. By measuring the rate of efflux from that of the inside of the vesicle to the ambient solution, allows researcher to better understand membrane permeability.

Vesicles can be formed with molecules and ions inside the vesicle by forming the vesicle with the desired molecule or ion present in the solution. Proteins can also be embedded into the membrane through solubilizing the desired proteins in the presence of detergents and attaching them to the phospholipids in which the liposome is formed. These provide researchers with a tool to examine various membrane protein functions.

Proteins

Type	*Description*	*Examples*
Integral proteins or transmembrane proteins	Span the membrane and have a hydrophilic cytosolic domain, which interacts with internal molecules, a hydrophobic membrane-spanning domain that anchors it within the cell membrane, and a hydrophilic extracellular domain that interacts with external molecules. The hydrophobic domain consists of one, multiple, or a combination of α-helices and β sheet protein motifs.	Ion channels, proton pumps, G protein-coupled receptor
Lipid anchored proteins	Covalently bound to single or multiple lipid molecules; hydrophobically insert into the cell membrane and anchor the protein. The protein itself is not in contact with the membrane.	G proteins
Peripheral proteins	Attached to integral membrane proteins, or associated with peripheral regions of the lipid bilayer. These proteins tend to have only temporary interactions with biological membranes, and once reacted, the molecule dissociates to carry on its work in the cytoplasm.	Some enzymes, some hormones

The cell membrane has large content of proteins, typically around 50% of membrane volume These proteins are important for cell because they are responsible for various biological activities. Approximately a third of the genes in yeast code specifically for them, and this number is even higher in multicellular organisms.

The cell membrane, being exposed to the outside environment, is an important site of cell–cell communication. As such, a large variety of protein receptors and identification proteins, such as antigens, are present on the surface of the membrane. Functions of membrane proteins can also include cell–cell contact, surface recognition, cytoskeleton contact, signalling, enzymatic activity, or transporting substances across the membrane.

Most membrane proteins must be inserted in some way into the membrane. For this to occur, an N-terminus "signal sequence" of amino acids directs proteins to the endoplasmic reticulum, which inserts the proteins into a lipid bilayer. Once inserted, the proteins are then transported to their final destination in vesicles, where the vesicle fuses with the target membrane.

Carbohydrates

Plasma membranes also contain carbohydrates, predominantly glycoproteins, but with some glycolipids (cerebrosides and gangliosides). For the most part, no glycosylation occurs on membranes within the cell; rather generally glycosylation occurs on the extracellular surface of the plasma membrane. The glycocalyx is an important feature in all cells, especially epithelia with microvilli. Recent data suggest the glycocalyx participates in cell adhesion, lymphocyte homing, and many others. The penultimate sugar is galactose and the terminal sugar is sialic acid, as the sugar backbone is modified in the golgi apparatus. Sialic acid carries a negative charge, providing an external barrier to charged particles.

Variation

The cell membrane has different lipid and protein compositions in distinct types of cells and may have therefore specific names for certain cell types:

- Sarcolemma in myocytes
- Oolemma in oocytes
- Axolemma in neuronal processes - axons
- Historically, the plasma membrane was also referred to as the plasmalemma

Permeability

The permeability of a membrane is the rate of passive diffusion of molecules through the membrane. These molecules are known as permeant molecules. Permeability depends mainly on the electric charge and polarity of the molecule and to a lesser extent the molar mass of the molecule. Due to the cell membrane's hydrophobic nature, small electrically neutral molecules pass through the membrane more easily than charged, large ones. The inability of charged molecules to pass through the cell membrane results in pH partition of substances throughout the fluid compartments of the body.

Membrane Transport

In cellular biology the term membrane transport refers to the collection of mechanisms that regulate the passage of solutes such as ions and small molecules through biological membranes, which are lipid bilayers that contain proteins embedded in them. The regulation of passage through the membrane is due to selective membrane permeability - a characteristic of biological membranes which allows them to separate substances of distinct chemical nature. In other words, they can be permeable to certain substances but not to others.

The movements of most solutes through the membrane are mediated by membrane transport proteins which are specialized to varying degrees in the transport of specific molecules. As the diversity and physiology of the distinct cells is highly related to their capacities to attract different external elements, it is postulated that there is a group of specific transport proteins for each cell type and for every specific physiological stage. This differential expression is regulated through the differential transcription of the genes coding for these proteins and its translation, for instance, through genetic-molecular mechanisms, but also at the cell biology level: the production of these proteins can be activated by cellular signalling pathways, at the biochemical level, or even by being situated in cytoplasmic vesicles.

Background

Thermodynamically the flow of substances from one compartment to another can occur in the direction of a concentration or electrochemical gradient or against it. If the exchange of substances occurs in the direction of the gradient, that is, in the direction of decreasing potential, there is no requirement for an input of energy from outside the system; if, however, the transport is against the gradient, it will require the input of energy, metabolic energy in this case. For example, a classic chemical mechanism for separation that does not require the addition of external energy is dialysis. In this system a semipermeable membrane separates two solutions of different concentration of the same solute. If the membrane allows the passage of water but not the solute the water will move into the compartment with the greatest solute concentration in order to establish an equilibrium in which the energy of the system is at a minimum. This takes place because the water moves from a high solvent concentration to a low one (in terms of the solute, the opposite occurs) and because the water is moving along a gradient there is no need for an external input of energy.

The nature of biological membranes, especially that of its lipids, is amphiphilic, as they form bilayers that contain an internal hydrophobic layer and an external hydrophilic layer. This structure makes transport possible by simple or passive diffusion, which consists of the diffusion of substances through the membrane without expending metabolic energy and without the aid of transport proteins. If the transported substance has a net electrical charge, it will move not only in response to a concentration gradient, but also to an electrochemical gradient due to the membrane potential.

Relative permeability of a phospholipid bilayer to various substances		
Type of substance	***Examples***	***Behaviour***
Gases	CO_2, N_2, O_2	Permeable
Small uncharged polar molecules	Urea, water, ethanol	Permeable, totally or partially
Large uncharged polar molecules	glucose, fructose	Not permeable
Ions	K^+, Na^+, Cl^-, HCO_3^-	Not permeable
Charged polar molecules	ATP, amino acids, glucose-6-phosphate	Not permeable

As few molecules are able to diffuse through a lipid membrane the majority of the transport processes involve transport proteins. These transmembrane proteins possess a large number of alpha helices immersed in the lipid matrix. In bacteria these proteins are present in the beta lamina form. This structure probably involves a conduit through hydrophilic protein environments that cause a disruption in the highly hydrophobic medium formed by the lipids. These proteins can be involved in transport in a number of ways: they act as pumps driven by ATP, that is, by metabolic energy, or as channels of facilitated diffusion.

Thermodynamics

A physiological process can only take place if it complies with basic thermodynamic principles. Membrane transport obeys physical laws that define its capabilities and therefore its biological utility.

A general principle of thermodynamics that governs the transfer of substances through membranes and other surfaces is that the exchange of free energy, ΔG, for the transport of a mole of a substance of concentration C_1 in a compartment to another compartment where it is present at C_2 is:

$$\Delta G = RT \log \frac{C_2}{C_1}.$$

Where C_2 is less than C_1 ΔG is negative, and the process is thermodynamically favourable. As the energy is transferred from one compartment to another, except where other factors intervene, an equilibrium will be reached where C_2=C_1, and where G=0. However, there are three circumstances under which this equilibrium will not be reached, circumstances which are vital for the *in vivo* functioning of biological membranes:

- The macromolecules on one side of the membrane can bond preferentially to a certain component of the membrane or chemically modify it. In this way, although the concentration of the solute may actually be different on both sides of the membrane, the availability of the solute is reduced in one of the compartments to such an extent that, for practical purposes, no gradient exists to drive transport.
- A membrane electrical potential can exist which can influence ion distribution. For example, for the transport of ions from the exterior to the interior, it is possible that:

$$\Delta G = RT \log \frac{C_{inside}}{C_{outside}} + ZF\Delta P$$

Where F is Faraday's constant and ΔP the membrane potential in volts. If ΔP is negative and Z is positive, the contribution of the term $ZF\Delta P$ to ΔG will be negative, that is, it will favour the transport of cations from the interior of the cell.

So, if the potential difference is maintained, the equilibrium state ΔG=0 will not correspond to an equimolar concentration of ions on both sides of the membrane.

- If a process with a negative ΔG is coupled to the transport process then the global ΔG will be modified. This situation is common in active transport and is described thus:

$$\Delta G = RT \log \frac{C_{inside}}{C_{outside}} + \Delta G^b$$

Where ΔG^b corresponds to a favourable thermodynamic reaction, such as the hydrolysis of ATP, or the co-transport of a compound that is moved in the direction of its gradient.

Transport Types

Passive Diffusion: As mentioned above, passive diffusion is a spontaneous phenomenon that increases the entropy of a system and decreases the free energy. The transport process is influenced by the characteristics of the transport substance and the nature of the bilayer. Membrane proteins (with the exception of channels - facilitated diffusion) are not involved in passive diffusion. The diffusion velocity of a pure phospholipid membrane will depend on:

- concentration gradient,
- hydrophobicity,
- size,
- charge, if the molecule has a net charge.

Active and co-transport

In active transport a solute is moved against a concentration or electrochemical gradient, in doing so the transport proteins involved consume metabolic energy, usually ATP. In primary active transport the hydrolysis of the energy provider (e.g. ATP) takes place directly in order to transport the solute in question, for instance, when the transport proteins are ATPase enzymes. Where the hydrolysis of the energy provider is indirect as is the case in secondary active transport, use is made of the energy stored in an electrochemical gradient. For example, in co-transport use is made of the gradients of certain solutes to transport a target compound against its gradient, causing the dissipation of the solute gradient. It may appear that, in this example, there is no energy use, but hydrolysis of the energy provider is required to establish the gradient of the solute transported along with the target compound. The gradient of the co-transported solute will be generated through the use of certain types of proteins called biochemical pumps.

The discovery of the existence of this type of transporter protein came from the study of the kinetics of cross-membrane molecule transport. For certain solutes it was noted that the transport velocity reached a plateau at a particular concentration above which there was no significant increase in uptake rate, indicating a log curve type response. This was interpreted as showing that transport was mediated by the formation of a substrate-transporter complex, which is conceptually the same as the enzyme-substrate complex of enzyme kinetics. Therefore, each transport protein has an affinity constant for a solute that is equal to the concentration of the solute when the transport velocity is half its maximum value. This is equivalent in the case of an enzyme to the Michaelis-Menten constant. Some important features of active transport in addition to its ability to intervene even against a gradient, its kinetics and the use of ATP, are its high selectivity and ease of selective pharmacological inhibition

Transporter proteins

A transport protein can move various ions and molecules, they are distinguished according to their directionality:

- antiporter: (also called exchanger or counter-transporter) transport proteins which transport a molecule against its gradient and at the same time displaces one or more ions along its gradient, both gradients being opposite,
- symporter: transport proteins which move a molecule against its gradient while displacing one or more different ions along their gradient which is in the same direction as that of the transported molecule.

Both can be referred to as co-transporters.

Pumps

A pump is a protein that hydrolyses ATP in order to transport a particular solute through a membrane in order to generate an electrochemical gradient to confer certain membrane potential characteristics on it. This gradient is of interest as an indicator of the state of the cell through parameters such as the Nernst potential. In terms of membrane transport the gradient is of interest as it contributes to increased system entropy in the co-transport of substances against their gradient. One of the most important pumps in animal cells is the sodium potassium pump, that operates through the following mechanism:

1. binding of three Na^+ ions to their active sites on the pump which are bound to ATP.
2. ATP is hydrolyzed leading to phosphorylation of the cytoplasmic side of the pump, this induces a structure change in the protein. The phosphorylation is caused by the transfer of the terminal group of ATP to a residue of aspartate in the transport protein and the subsequent release of ADP.
3. the structure change in the pump exposes the Na^+ to the exterior. The phosphorylated form of the pump has a low affinity for Na^+ ions so they are released.
4. once the Na^+ ions are liberated, the pump binds two molecules of K^+ to their respective bonding sites on the extracellular face

of the transport protein.This causes the dephosphorylation of the pump, reverting it to its previous conformational state, transporting the K^+ ions into the cell.

5. The unphosphorylated form of the pump has a higher affinity for Na^+ ions than K^+ ions, so the two bound K^+ ions are released into the cytosol. ATP binds, and the process starts again.

Membrane Selectivity

As the main characteristic of transport through a biological membrane is its selectivity and its subsequent behaviour as a barrier for certain substances, the underlying physiology of the phenomenon has been studied extensively. Investigation into membrane selectivity have classically been divided into those relating to electrolytes and non-electrolytes.

Electrolyte Selectivity

The ionic channels define an internal diameter that permits the passage of small ions that is related to various characteristics of the ions that could potentially be transported. As the size of the ion is related to its chemical species, it could be assumed *a priori* that a channel whose pore diameter was sufficient to allow the passage of one ion would also allow the transfer of others of smaller size, however, this does not occur in the majority of cases. There are two characteristics alongside size that are important in the determination of the selectivity of the membrane pores: the facility for dehydration and the interaction of the ion with the internal charges of the pore.

In order for an ion to pass through a pore it must dissociate itself from the water molecules that cover it in successive layers of solvation. The tendency to dehydrate, or the facility to do this, is related to the size of the ion: larger ions can do it more easily that the smaller ions, so that a pore with weak polar centres will preferentially allow passage of larger ions over the smaller ones.

When the interior of the channel is composed of polar groups from the side chains of the component amino acids, the interaction of a dehydrated ion with these centres can be more important than the facility for dehydration in conferring the specificity of the channel. For example, a channel made up of histidines and arginines, with positively charged groups, will selectively repel ions of the same polarity, but will facilitate the passage of negatively charged ions. Also, in this case, the smallest ions will be able to interact more closely due to the spatial arrangement of the molecule (stericity), which greatly increases the charge-charge interactions and therefore exaggerates the effect.

Non-electrolyte Selectivity

Non-electrolytes, substances that generally are hydrophobic and lipophylic, usually pass through the membrane by dissolution in the lipid bilayer, and therefore, by passive diffusion. For those non-electrolytes whose transport through the membrane is mediated by a transport protein the ability to diffuse is, generally, dependent on the partition coefficient K. Partially charged non-electrolytes, that are more or less polar, such as ethanol, methanol or urea, are able to pass through the membrane through aqueous channels immersed in the membrane. It is interesting to note that there is no effective regulation mechanism that limits this transport, which indicates an intrinsic vulnerability of the cells to the penetration of these molecules.

Creation of Membrane Transport Proteins

There are several databases which attempt to construct phylogenetic trees detailing the creation of transporter proteins. One such resource is the Transporter Classification database

Photosynthesis

Photosynthesis is a process used by plants and other organisms to convert light energy, normally from the sun, into chemical energy that can be used to fuel the organisms' activities. Carbohydrates, such as sugars, are synthesized from carbon dioxide and water. Oxygen is also released, mostly as a waste product. Most plants, most algae, and cyanobacteria perform the process of photosynthesis, and are called photoautotrophs. Photosynthesis maintains atmospheric oxygen levels and supplies all of the organic compounds and most of the energy necessary for all life on Earth.

Although photosynthesis is performed differently by different species, the process always begins when energy from light is absorbed by proteins called reaction centres that contain green chlorophyll pigments. In plants, these proteins are held inside organelles called chloroplasts, which are most abundant in leaf cells, while in bacteria they are embedded in the plasma membrane. In these light-dependent reactions, some energy is used to strip electrons from suitable substances such as water, producing oxygen gas. Furthermore, two further compounds are generated: reduced nicotinamide adenine dinucleotide phosphate (NADPH) and adenosine triphosphate (ATP), the "energy currency" of cells.

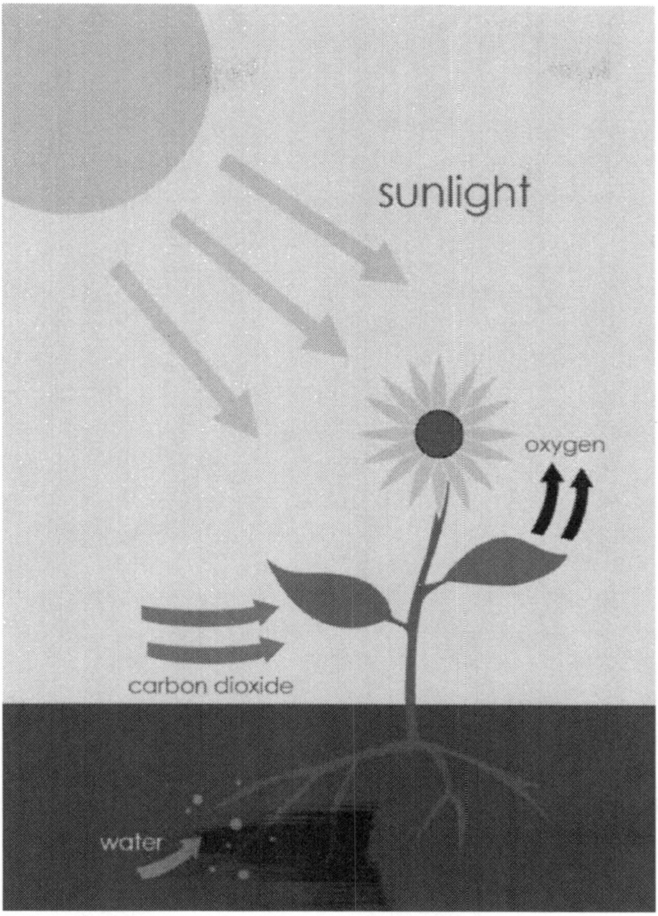

Figure: *Schematic of photosynthesis in plants. The carbohydrates produced are stored in or used by the plant.*

In plants, algae and cyanobacteria, sugars are produced by a subsequent sequence of light-independent reactions called the Calvin cycle, but some bacteria use different mechanisms, such as the reverse Krebs cycle. In the Calvin cycle, atmospheric carbon dioxide is incorporated into already existing organic carbon compounds, such as ribulose bisphosphate (RuBP). Using the ATP and NADPH produced by the light-dependent reactions, the resulting compounds are then reduced and removed to form further carbohydrates such as glucose.

The first photosynthetic organisms probably evolved early in the evolutionary history of life and most likely used reducing agents such as hydrogen or hydrogen sulfide as sources of electrons, rather than water. Cyanobacteria appeared later, and the excess oxygen they produced contributed to the oxygen catastrophe, which rendered the evolution of complex life possible. Today, the average rate of energy capture by photosynthesis globally is approximately 130 terawatts, which is about six times larger than the current power consumption of human civilization. Photosynthetic organisms also convert around 100–115 thousand million metric tonnes of carbon into biomass per year.

Overview

Photosynthetic organisms are photoautotrophs, which means that they are able to synthesize food directly from carbon dioxide and water using energy from light. However, not all organisms that use light as a source of energy carry out photosynthesis, since *photoheterotrophs* use organic compounds, rather than carbon dioxide, as a source of carbon. In plants, algae and cyanobacteria, photosynthesis releases oxygen. This is called *oxygenic photosynthesis*. Although there are some differences between oxygenic photosynthesis in plants, algae, and cyanobacteria, the overall process is quite similar in these organisms. However, there are some types of bacteria that carry out anoxygenic photosynthesis, which consumes carbon dioxide but does not release oxygen.

Carbon dioxide is converted into sugars in a process called carbon fixation. Carbon fixation is an endothermic redox reaction, so photosynthesis needs to supply both a source of energy to drive this process, and the electrons needed to convert carbon dioxide into a carbohydrate. This addition of the electrons is a reduction reaction. In general outline and in effect, photosynthesis is the opposite of cellular respiration, in which glucose and other compounds are oxidized to produce carbon dioxide and water, and to release exothermic chemical energy to drive the organism's metabolism. However, the two processes take place through a different sequence of chemical reactions and in different cellular compartments.

The general equation for photosynthesis is therefore:

$$2n\ CO_2 + 2n\ DH_2 + \text{photons} \rightarrow 2(CH_2O)_n + 2n\ DO$$

Carbon dioxide + electron donor + light energy → carbohydrate + oxidized electron donor

In *oxygenic* photosynthesis water is the electron donor and, since its hydrolysis releases oxygen, the equation for this process is:

$$2n\ CO_2 + 4n\ H_2O + \text{photons} \rightarrow 2(CH_2O)_n + 2n\ O_2 + 2n\ H_2O$$

carbon dioxide + water + light energy → carbohydrate + oxygen + water

Often 2n water molecules are cancelled on both sides, yielding:

$2n\ CO_2 + 2n\ H_2O + \text{photons} \rightarrow 2(CH_2O)_n + 2n\ O_2$

carbon dioxide + water + light energy → carbohydrate + oxygen

Other processes substitute other compounds (such as arsenite) for water in the electron-supply role; for example some microbes use sunlight to oxidize arsenite to arsenate: The equation for this reaction is:

$CO_2 + (AsO_3^{3-}) + \text{photons} \rightarrow (AsO_4^{3-}) + CO$

carbon dioxide + arsenite + light energy → arsenate + carbon monoxide (used to build other compounds in subsequent reactions)

Photosynthesis occurs in two stages. In the first stage, *light-dependent reactions* or *light reactions* capture the energy of light and use it to make the energy-storage molecules ATP and NADPH. During the second stage, the *light-independent reactions* use these products to capture and reduce carbon dioxide.

Most organisms that utilize photosynthesis to produce oxygen use visible light to do so, although at least three use shortwave infrared or, more specifically, far-red radiation.

Photosynthetic Membranes and Organelles

In photosynthetic bacteria, the proteins that gather light for photosynthesis are embedded within cell membranes, which is the simplest configuration these proteins are arranged. However, this membrane may be tightly folded into cylindrical sheets called thylakoids, or bunched up into round vesicles called *intracytoplasmic membranes*. These structures can fill most of the interior of a cell, giving the membrane a very large surface area and therefore increasing the amount of light that the bacteria can absorb.

In plants and algae, photosynthesis takes place in organelles called chloroplasts. A typical plant cell contains about 10 to 100 chloroplasts. The chloroplast is enclosed by a membrane. This membrane is composed of a phospholipid inner membrane, a phospholipid outer membrane, and an intermembrane space between them. Within the membrane is an aqueous fluid called the stroma. The stroma contains stacks (grana) of thylakoids, which are the site of photosynthesis. The thylakoids are flattened disks, bounded by a membrane with a lumen or thylakoid space within it. The site of photosynthesis is the thylakoid membrane, which contains integral and peripheral membrane protein complexes, including the pigments that absorb light energy, which form the photosystems.

Plants absorb light primarily using the pigment chlorophyll, which is the reason that most plants have a green colour. Besides chlorophyll, plants also use pigments such as carotenes and xanthophylls. Algae also use chlorophyll, but various other pigments are present as phycocyanin, carotenes, and xanthophylls in green algae, phycoerythrin in red algae (rhodophytes) and fucoxanthin in brown algae and diatoms resulting in a wide variety of colours.

These pigments are embedded in plants and algae in special antenna-proteins. In such proteins all the pigments are ordered to work well together. Such a protein is also called a light-harvesting complex.

Although all cells in the green parts of a plant have chloroplasts, most of the energy is captured in the leaves, except in certain species adapted to conditions of strong sunlight and aridity, such as many Euphorbia and Cactus species, whose main photosynthetic organs are their stems. The cells in the interior tissues of a leaf, called the mesophyll, can contain between 450,000 and 800,000 chloroplasts for every square millimeter of leaf. The surface of the leaf is uniformly coated with a water-resistant waxy cuticle that protects the leaf from excessive evaporation of water and decreases the absorption of ultraviolet or blue light to reduce heating. The transparent epidermis layer allows light to pass through to the palisade mesophyll cells where most of the photosynthesis takes place.

Light Reactions

In the light reactions, one molecule of the pigment chlorophyll absorbs one photon and loses one electron. This electron is passed to a modified form of chlorophyll called pheophytin, which passes the electron to a quinone molecule, allowing the start of a flow of electrons down an electron transport chain that leads to the ultimate reduction of NADP to NADPH. In addition, this creates a proton gradient across the chloroplast membrane; its dissipation is used by ATP synthase for the concomitant synthesis of ATP. The chlorophyll molecule regains the lost electron from a water molecule through a process called photolysis, which releases a dioxygen (O_2) molecule. The overall equation for the light-dependent reactions under the conditions of non-cyclic electron flow in green plants is:

$$2\ H_2O + 2\ NADP^+ + 3\ ADP + 3\ P_i + \text{light} \rightarrow 2\ NADPH + 2\ H^+ + 3\ ATP + O_2$$

Not all wavelengths of light can support photosynthesis. The photosynthetic action spectrum depends on the type of accessory pigments present. For example, in green plants, the action spectrum resembles the absorption spectrum for chlorophylls and carotenoids with peaks for violet-blue and red light. In red algae, the action spectrum overlaps with the absorption spectrum of phycobilins for red blue-green light, which allows these algae to grow in deeper waters that filter out the longer wavelengths used by green plants. The non-absorbed part of the light spectrum is what gives photosynthetic organisms their colour (e.g., green plants, red algae, purple bacteria) and is the least effective for photosynthesis in the respective organisms.

Z Scheme

In plants, light-dependent reactions occur in the thylakoid membranes of the chloroplasts and use light energy to synthesize ATP and NADPH. The light-dependent reaction has two forms: cyclic and non-cyclic. In the non-cyclic reaction, the photons are captured in the light-harvesting antenna complexes of photosystem II by chlorophyll and other accessory pigments. When a chlorophyll molecule at the core of the photosystem II reaction centre obtains sufficient excitation energy from the adjacent antenna pigments, an electron is transferred to the primary electron-acceptor molecule, pheophytin, through a process called photoinduced charge separation.

These electrons are shuttled through an electron transport chain, the so-called *Z-scheme* shown in the diagram, that initially functions to generate a chemiosmotic potential across the membrane. An ATP synthase enzyme uses the chemiosmotic potential to make ATP during photophosphorylation, whereas NADPH is a product of the terminal redox reaction in the *Z-scheme*. The electron enters a chlorophyll molecule in Photosystem I. The electron is excited due to the light absorbed by the photosystem.

A second electron carrier accepts the electron, which again is passed down lowering energies of electron acceptors. The energy created by the electron acceptors is used to move hydrogen ions across the thylakoid membrane into the lumen. The electron is used to reduce the co-enzyme NADP, which has functions in the light-independent reaction. The cyclic reaction is similar to that of the non-cyclic, but differs in the form that it generates only ATP, and no reduced NADP (NADPH) is created. The cyclic reaction takes place only at photosystem I. Once the electron is displaced from the photosystem, the electron is passed down the electron acceptor molecules and returns to photosystem I, from where it was emitted, hence the name *cyclic reaction*.

Water Photolysis

The NADPH is the main reducing agent in chloroplasts, providing a source of energetic electrons to other reactions. Its production leaves chlorophyll with a deficit of electrons (oxidized), which must be obtained from some other reducing agent. The excited electrons lost from chlorophyll in photosystem I are replaced from the electron transport chain by plastocyanin. However, since photosystem II includes the first steps of the *Z-scheme*, an external source of electrons is required to reduce its oxidized chlorophyll *a* molecules. The source of electrons in green-plant and cyanobacterial photosynthesis is water. Two water molecules are oxidized by four successive charge-separation reactions by photosystem II to yield a molecule of diatomic oxygen and four hydrogen ions; the electron yielded in each step is transferred to a redox-active tyrosine residue that then reduces the photoxidized paired-chlorophyll *a* species called P680 that serves as the primary (light-driven) electron donor in the photosystem II reaction centre. The oxidation of water is catalyzed in photosystem II by a redox-active structure that contains four manganese ions and a calcium ion; this oxygen-evolving complex binds two water molecules and stores the four oxidizing equivalents that are required to drive the water-oxidizing reaction. Photosystem II is the only known biological enzyme that carries out this oxidation of water. The hydrogen ions contribute to the transmembrane chemiosmotic potential that leads to ATP synthesis. Oxygen is a waste product of light-dependent reactions, but the majority of organisms on Earth use oxygen for cellular respiration, including photosynthetic organisms.

Light-independent Reactions

Calvin Cycle

In the light-independent (or "dark") reactions, the enzyme RuBisCO captures CO_2 from the atmosphere and in a process that requires the newly formed NADPH, called the Calvin-Benson Cycle, releases three-carbon sugars, which are later combined to form sucrose and starch. The overall equation for the light-independent reactions in green plants is:

$$3\ CO_2 + 9\ ATP + 6\ NADPH + 6\ H^+ \rightarrow C_3H_6O_3\text{-phosphate} + 9\ ADP + 8\ P_i + 6\ NADP^+ + 3\ H_2O$$

To be more specific, carbon fixation produces an intermediate product, which is then converted to the final carbohydrate products. The carbon skeletons produced by photosynthesis are then variously used to form other organic compounds, such as the building material cellulose, as precursors for lipid and amino acid biosynthesis, or as a fuel in cellular respiration. The latter occurs not only in plants but also in animals when the energy from plants gets passed through a food chain.

The fixation or reduction of carbon dioxide is a process in which carbon dioxide combines with a five-carbon sugar, ribulose 1,5-bisphosphate (RuBP), to yield two molecules of a three-carbon compound, glycerate 3-phosphate (GP), also known as 3-phosphoglycerate (PGA). GP, in the presence of ATP and NADPH from the light-dependent stages, is reduced to glyceraldehyde 3-phosphate (G3P). This product is also referred to as 3-phosphoglyceraldehyde (PGAL) or even as triose phosphate. Triose is a 3-carbon sugar. Most (5 out of 6 molecules) of the G3P produced is used to regenerate RuBP so the process can continue. The 1 out of 6 molecules of the triose phosphates not "recycled" often condense to form hexose phosphates, which ultimately yield sucrose, starch and cellulose. The sugars produced during carbon metabolism yield carbon skeletons that can be used for other metabolic reactions like the production of amino acids and lipids.

Carbon Concentrating Mechanisms

On Land: In hot and dry conditions, plants close their stomata to prevent the loss of water. Under these conditions, CO_2 will decrease, and oxygen gas, produced by the light reactions of photosynthesis, will decrease in the stem, not leaves, causing an increase of photorespiration by the oxygenase activity of ribulose-1,5-bisphosphate carboxylase/oxygenase and decrease in carbon fixation. Some plants have evolved mechanisms to increase the CO_2 concentration in the leaves under these conditions.

C_4 plants chemically fix carbon dioxide in the cells of the mesophyll by adding it to the three-carbon molecule phosphoenolpyruvate (PEP), a reaction catalyzed by an enzyme called PEP carboxylase, creating the four-carbon organic acid oxaloacetic acid. Oxaloacetic acid or malate synthesized by this process is then translocated to specialized bundle sheath cells where the enzyme RuBisCO and other Calvin cycle enzymes are located, and where CO_2 released by decarboxylation of the four-carbon acids is then fixed by RuBisCO activity to the three-carbon sugar 3-phosphoglyceric acids. The physical separation of RuBisCO from the oxygen-generating light reactions reduces photorespiration and increases CO_2 fixation and, thus, photosynthetic capacity of the leaf. C_4 plants can produce more sugar than C_3 plants in conditions of high light and temperature.

Many important crop plants are C_4 plants, including maize, sorghum, sugarcane, and millet. Plants that do not use PEP-carboxylase in carbon fixation are called C_3 plants because the primary carboxylation reaction, catalyzed by RuBisCO, produces the three-carbon sugar 3-phosphoglyceric acids directly in the Calvin-Benson cycle.

Over 90% of plants use C_3 carbon fixation, compared to 3% that use C_4 carbon fixation.; however, the fact that C_4 has evolved in over 60 plant lineages makes it a striking example of convergent evolution.

Xerophytes, such as cacti and most succulents, also use PEP carboxylase to capture carbon dioxide in a process called Crassulacean acid metabolism (CAM). In contrast to C_4 metabolism, which *physically* separates the CO_2 fixation to PEP from the Calvin cycle, CAM *temporally* separates these two processes.

CAM plants have a different leaf anatomy from C_3 plants, and fix the CO_2 at night, when their stomata are open. CAM plants store the CO_2 mostly in the form of malic acid via carboxylation of phosphoenolpyruvate to oxaloacetate, which is then reduced to malate. Decarboxylation of malate during the day releases CO_2 inside the leaves, thus allowing carbon fixation to 3-phosphoglycerate by RuBisCO. Sixteen thousand species of plants use CAM.

In Water

Cyanobacteria possess carboxysomes, which increase the concentration of CO_2 around RuBisCO to increase the rate of photosynthesis. An enzyme, carbonic anhydrase, located within the carboxysome releases CO_2 from the dissolved hydrocarbonate ions (HCO_3^-). Before the CO_2 diffuses out it is quickly sponged up by RuBisCO, which is concentrated within the carboxysomes.

HCO_3^- ions are made from CO_2 outside the cell by another carbonic anhydrase and are actively pumped into the cell by a membrane protein. They cannot cross the membrane as they are charged, and within the cytosol they turn back into CO_2 very slowly without the help of carbonic anhydrase.

This causes the HCO_3^- ions to accumulate within the cell from where they diffuse into the carboxysomes. Pyrenoids in algae and hornworts also act to concentrate CO_2 around rubisco.

Order and Kinetics

The overall process of photosynthesis takes place in four stages:

Stage	*Description*	*Time scale*
1	Energy transfer in antenna chlorophyll (thylakoid membranes)	femtosecond to picosecond
2	Transfer of electrons in photochemical reactions (thylakoid membranes)	picosecond to nanosecond
3	Electron transport chain and ATP synthesis (thylakoid membranes)	microsecond to millisecond
4	Carbon fixation and export of stable products	millisecond to second

Efficiency

Plants usually convert light into chemical energy with a photosynthetic efficiency of 3–6%. Absorbed light that is unconverted is dissipated primarily as heat, with a small fraction (1-2%) re-emitted as chlorophyll fluorescence at longer (redder) wavelengths.

Actual plants' photosynthetic efficiency varies with the frequency of the light being converted, light intensity, temperature and proportion of carbon dioxide in the atmosphere, and can vary from 0.1% to 8%. By comparison, solar panels convert light into electric energy at an efficiency of approximately 6–20% for mass-produced panels, and above 40% in laboratory devices.

Photosynthesis measurement systems are not designed to directly measure the amount of light absorbed by the leaf. Nevertheless, the light response curves that systems like the LCpro-SD produce, do allow comparisons in photosynthetic efficiency between plants.

Evolution

Early photosynthetic systems, such as those from green and purple sulphur and green and purple nonsulphur bacteria, are thought to have been anoxygenic, using various molecules as electron donors. Green and purple sulphur bacteria are thought to have used hydrogen and sulphur as an electron donor. Green nonsulphur bacteria used various amino and other organic acids. Purple nonsulphur bacteria used a variety of nonspecific organic molecules. The use of these molecules is consistent with the geological evidence that the atmosphere was highly reduced at that time.

Fossils of what are thought to be filamentous photosynthetic organisms have been dated at 3.4 billion years old.

The main source of oxygen in the atmosphere is oxygenic photosynthesis, and its first appearance is sometimes referred to as the oxygen catastrophe. Geological evidence suggests that oxygenic photosynthesis, such as that in cyanobacteria, became important during the Paleoproterozoic era around 2 billion years ago. Modern photosynthesis in plants and most photosynthetic prokaryotes is oxygenic. Oxygenic photosynthesis uses water as an electron donor, which is oxidized to molecular oxygen (O_2) in the photosynthetic reaction centre.

Symbiosis and the Origin of Chloroplasts

Several groups of animals have formed symbiotic relationships with photosynthetic algae. These are most common in corals, sponges and sea anemones. It is presumed that this is due to the particularly simple body plans and large surface areas of these animals compared to their volumes. In addition, a few marine mollusks *Elysia viridis* and *Elysia chlorotica* also maintain a symbiotic relationship with chloroplasts they capture from the algae in their diet and then store in their bodies. This allows the mollusks to survive solely by photosynthesis for several months at a time. Some of the genes from the plant cell nucleus have even been transferred to the slugs, so that the chloroplasts can be supplied with proteins that they need to survive.

An even closer form of symbiosis may explain the origin of chloroplasts. Chloroplasts have many similarities with photosynthetic bacteria, including a circular chromosome, prokaryotic-type ribosomes, and similar proteins in the photosynthetic reaction centre. The endosymbiotic theory suggests that photosynthetic bacteria were acquired (by endocytosis) by early eukaryotic cells to form the first plant cells. Therefore, chloroplasts may be photosynthetic bacteria that adapted to life inside plant cells. Like mitochondria, chloroplasts still possess their own DNA, separate from the nuclear DNA of their plant host cells and the genes in this chloroplast DNA resemble those in cyanobacteria. DNA in chloroplasts codes for redox proteins such as photosynthetic reaction centres. The CoRR Hypothesis proposes that this Co-location is required for Redox Regulation.

Cyanobacteria and the Evolution of Photosynthesis

The biochemical capacity to use water as the source for electrons in photosynthesis evolved once, in a common ancestor of extant cyanobacteria. The geological record indicates that this transforming event took place early in Earth's history, at least 2450–2320 million years ago (Ma), and, it is speculated, much earlier. Available evidence from geobiological studies of Archean (>2500 Ma) sedimentary rocks indicates that life existed 3500 Ma, but the question of when oxygenic photosynthesis evolved is still unanswered. A clear paleontological window on cyanobacterial evolution opened about 2000 Ma, revealing an already-diverse biota of blue-greens. Cyanobacteria remained principal primary producers throughout the Proterozoic Eon (2500–543 Ma), in part because the redox structure of the oceans favoured photoautotrophs capable of nitrogen fixation. Green algae joined blue-greens as major primary producers on continental shelves near the end of the Proterozoic, but only with the Mesozoic (251–65 Ma) radiations of dinoflagellates, coccolithophorids, and diatoms did primary production in marine shelf waters take modern form. Cyanobacteria remain critical to marine ecosystems as primary producers in oceanic gyres, as agents of biological nitrogen fixation, and, in modified form, as the plastids of marine algae.

A 2010 study by researchers at Tel Aviv University discovered that the Oriental hornet (*Vespa orientalis*) converts sunlight into electric power using a pigment called xanthopterin. This is the first scientific evidence of a member of the animal kingdom engaging in photosynthesis.

Discovery

Although some of the steps in photosynthesis are still not completely understood, the overall photosynthetic equation has been known since the 19th century.

Jan van Helmont began the research of the process in the mid-17th century when he carefully measured the mass of the soil used by a plant and the mass of the plant as it grew. After noticing that the soil mass changed very little, he hypothesized that the mass of the growing plant must come from the water, the only substance he added to the potted plant. His hypothesis was partially accurate — much of the gained mass also comes from carbon dioxide as well as water. However, this was a signalling point to the idea that the bulk of a plant's biomass comes from the inputs of photosynthesis, not the soil itself.

Joseph Priestley, a chemist and minister, discovered that, when he isolated a volume of air under an inverted jar, and burned a candle in it, the candle would burn out very quickly, much before it ran out of wax. He further discovered that a mouse could similarly "injure" air. He then showed that the air that had been "injured" by the candle and the mouse could be restored by a plant.

In 1778, Jan Ingenhousz, court physician to the Austrian Empress, repeated Priestley's experiments. He discovered that it was the influence of sunlight on the plant that could cause it to revive a mouse in a matter of hours.

In 1796, Jean Senebier, a Swiss pastor, botanist, and naturalist, demonstrated that green plants consume carbon dioxide and release oxygen under the influence of light. Soon afterward, Nicolas-Théodore de Saussure showed that the increase in mass of the plant as it grows could not be due only to uptake of CO_2 but also to the incorporation of water. Thus, the basic reaction by which photosynthesis is used to produce food (such as glucose) was outlined.

Cornelis Van Niel made key discoveries explaining the chemistry of photosynthesis. By studying purple sulphur bacteria and green bacteria he was the first scientist to demonstrate that photosynthesis is a light-dependent redox reaction, in which hydrogen reduces carbon dioxide. Robert Emerson discovered two light reactions by testing plant productivity using different wavelengths of light. With the red alone, the light reactions were suppressed. When blue and red were combined, the output was much more substantial. Thus, there were two photosystems, one absorbing up to 600 nm wavelengths, the other up to 700 nm. The former is known as PSII, the latter is PSI. PSI contains only chlorophyll a, PSII contains primarily chlorophyll a with most of the available chlorophyll b, among other pigment. These include phycobilins, which are the red and blue pigments of red and blue algae respectively, and fucoxanthol for brown algae and diatoms. The process is most productive when absorption of quanta are equal in both the PSII and PSI, assuring that input energy from the antenna complex is divided between the PSI and PSII system, which in turn powers the photochemistry.

Robert Hill thought that a complex of reactions consisting of an intermediate to cytochrome b_6 (now a plastoquinone), another is from cytochrome f to a step in the carbohydrate-generating mechanisms. These are linked by plastoquinone, which does require energy to reduce cytochrome f for it is a sufficient reductant. Further experiments to prove that the oxygen developed during the photosynthesis of green plants came from water, were performed by Hill in 1937 and 1939. He showed that isolated chloroplasts give off oxygen in the presence of unnatural reducing agents like iron oxalate, ferricyanide or benzoquinone after exposure to light. The Hill reaction is as follows:

$2\ H_2O + 2\ A +$ (light, chloroplasts) $\rightarrow 2\ AH_2 + O_2$ where A is the electron acceptor. Therefore, in light, the electron acceptor is reduced and oxygen is evolved.

Samuel Ruben and Martin Kamen used radioactive isotopes to determine that the oxygen liberated in photosynthesis came from the water. Melvin Calvin and Andrew Benson, along with James Bassham, elucidated the path of carbon assimilation (the photosynthetic carbon reduction cycle) in plants. The carbon reduction cycle is known as the Calvin cycle, which ignores the contribution of Bassham and Benson. Many scientists refer to the cycle as the Calvin-Benson Cycle, Benson-Calvin, and some even call it the Calvin-Benson-Bassham (or CBB) Cycle.

Nobel Prize-winning scientist Rudolph A. Marcus was able to discover the function and significance of the electron transport chain.

Otto Heinrich Warburg and Dean Burk discovered the I-quantum photosynthesis reaction that splits the CO_2, activated by the respiration.

Louis N.M. Duysens and Jan Amesz discovered that chlorophyll a will absorb one light, oxidize cytochrome f, chlorophyll a (and other pigments) will absorb another light, but will reduce this same oxidized cytochrome, stating the two light reactions are in series.

Development of the Concept

In 1893, Charles Reid Barnes proposed two terms, *photosyntax* and *photosynthesis*, for the biological process of *synthesis of complex carbon compounds out of carbonic acid, in the presence of chlorophyll, under the influence of light.* Over time, the term *photosynthesis* came into common usage as the term of choice. Later discovery of anoxygenic photosynthetic bacteria and photophosphorylation necessitated redefinition of the term.

Factors

There are three main factors affecting photosynthesis and several corollary factors. The three main are:

- Light irradiance and wavelength
- Carbon dioxide concentration
- Temperature.

Light Intensity (Irradiance), Wavelength and Temperature

In the early 20th century, Frederick Blackman and Gabrielle Matthaei investigated the effects of light intensity (irradiance) and temperature on the rate of carbon assimilation.

- At constant temperature, the rate of carbon assimilation varies with irradiance, initially increasing as the irradiance increases.

However, at higher irradiance, this relationship no longer holds and the rate of carbon assimilation reaches a plateau.

- At constant irradiance, the rate of carbon assimilation increases as the temperature is increased over a limited range. This effect is seen only at high irradiance levels. At low irradiance, increasing the temperature has little influence on the rate of carbon assimilation.

These two experiments illustrate vital points: First, from research it is known that, in general, photochemical reactions are not affected by temperature. However, these experiments clearly show that temperature affects the rate of carbon assimilation, so there must be two sets of reactions in the full process of carbon assimilation. These are, of course, the light-dependent 'photochemical' stage and the light-independent, temperature-dependent stage. Second, Blackman's experiments illustrate the concept of limiting factors. Another limiting factor is the wavelength of light. Cyanobacteria, which reside several metres underwater, cannot receive the correct wavelengths required to cause photoinduced charge separation in conventional photosynthetic pigments. To combat this problem, a series of proteins with different pigments surround the reaction centre. This unit is called a phycobilisome.

Carbon Dioxide Levels and Photorespiration

As carbon dioxide concentrations rise, the rate at which sugars are made by the light-independent reactions increases until limited by other factors. RuBisCO, the enzyme that captures carbon dioxide in the light-independent reactions, has a binding affinity for both carbon dioxide and oxygen. When the concentration of carbon dioxide is high, RuBisCO will fix carbon dioxide. However, if the carbon dioxide concentration is low, RuBisCO will bind oxygen instead of carbon dioxide. This process, called photorespiration, uses energy, but does not produce sugars.

RuBisCO oxygenase activity is disadvantageous to plants for several reasons:

1. One product of oxygenase activity is phosphoglycolate (2 carbon) instead of 3-phosphoglycerate (3 carbon). Phosphoglycolate cannot be metabolized by the Calvin-Benson cycle and represents carbon lost from the cycle. A high oxygenase activity, therefore, drains the sugars that are required to recycle ribulose 5-bisphosphate and for the continuation of the Calvin-Benson cycle.

2. Phosphoglycolate is quickly metabolized to glycolate that is toxic to a plant at a high concentration; it inhibits photosynthesis.
3. Salvaging glycolate is an energetically expensive process that uses the glycolate pathway, and only 75% of the carbon is returned to the Calvin-Benson cycle as 3-phosphoglycerate. The reactions also produce ammonia (NH_3), which is able to diffuse out of the plant, leading to a loss of nitrogen.

A highly simplified summary is: 2 glycolate + ATP '! 3-phosphoglycerate + carbon dioxide + ADP + NH_3

The salvaging pathway for the products of RuBisCO oxygenase activity is more commonly known as photorespiration, since it is characterized by light-dependent oxygen consumption and the release of carbon dioxide.

Plant Tissue Culture

Plant tissue culture is a collection of techniques used to maintain or grow plant cells, tissues or organs under sterile conditions on a nutrient culture medium of known composition. Plant tissue culture is widely used to produce clones of a plant in a method known as micropropagation. Different techniques in plant tissue culture may offer certain advantages over traditional methods of propagation, including:

- The production of exact copies of plants that produce particularly good flowers, fruits, or have other desirable traits.
- To quickly produce mature plants.
- The production of multiples of plants in the absence of seeds or necessary pollinators to produce seeds.
- The regeneration of whole plants from plant cells that have been genetically modified.
- The production of plants in sterile containers that allows them to be moved with greatly reduced chances of transmitting diseases, pests, and pathogens.
- The production of plants from seeds that otherwise have very low chances of germinating and growing, i.e.: orchids and *Nepenthes*.
- To clean particular plants of viral and other infections and to quickly multiply these plants as 'cleaned stock' for horticulture and agriculture.

Plant tissue culture relies on the fact that many plant cells have the ability to regenerate a whole plant (totipotency). Single cells, plant cells without cell walls (protoplasts), pieces of leaves, stems or roots can often be used to generate a new plant on culture media given the required nutrients and plant hormones.

Techniques

Modern plant tissue culture is performed under aseptic conditions under HEPA filtered air provided by a laminar flow cabinet. Living plant materials from the environment are naturally contaminated on their surfaces (and sometimes interiors) with microorganisms, so surface sterilization of starting material (explants) in chemical solutions (usually alcohol and sodium or calcium hypochlorite or mercuric chloride is required.

Mercuric chloride is seldom used as a plant sterilant today, unless other sterilizing agents are found to be ineffective, as it is dangerous to use, and is difficult to dispose of. Explants are then usually placed on the surface of a solid culture medium, but are sometimes placed directly into a liquid medium, particularly when cell suspension cultures are desired. Solid and liquid media are generally composed of inorganic salts plus a few organic nutrients, vitamins and plant hormones. Solid media are prepared from liquid media with the addition of a gelling agent, usually purified agar.

The composition of the medium, particularly the plant hormones and the nitrogen source (nitrate versus ammonium salts or amino acids) have profound effects on the morphology of the tissues that grow from the initial explant. For example, an excess of auxin will often result in a proliferation of roots, while an excess of cytokinin may yield shoots.

A balance of both auxin and cytokinin will often produce an unorganised growth of cells, or callus, but the morphology of the outgrowth will depend on the plant species as well as the medium composition. As cultures grow, pieces are typically sliced off and transferred to new media (subcultured) to allow for growth or to alter the morphology of the culture. The skill and experience of the tissue culturist are important in judging which pieces to culture and which to discard.

As shoots emerge from a culture, they may be sliced off and rooted with auxin to produce plantlets which, when mature, can be transferred to potting soil for further growth in the greenhouse as normal plants.

Choice of Explant

The tissue obtained from a plant to be cultured is called an explant based on work with certain model systems particularly tobacco it has often been claimed that a totipotent explant can be grown from any part of the plant and may include portions of shoots, leaves, stems, flowers, roots and single, undifferentiated cells., however this has not been true for all plants. In many species explants of various organs vary in their rates of growth and regeneration, while some do not grow at all. The choice of explant material also determines if the plantlets developed via tissue culture are haploid or diploid. Also the risk of microbial contamination is increased with inappropriate explants.

The specific differences in the regeneration potential of different organs and explants have various explanations. The significant factors include differences in the stage of the cells in the cell cycle, the availability of or ability to transport endogenous growth regulators, and the metabolic capabilities of the cells. The most commonly used tissue explants are the meristematic ends of the plants like the stem tip, auxiliary bud tip and root tip. These tissues have high rates of cell division and either concentrate or produce required growth regulating substances including auxins and cytokinins.

The pathways through which whole plants are regenerated from cells and tissues or explants such as meristems broadly fall into three types:

1. The method in which explants that include a meristem (viz. the shoot tips or nodes) are grown on appropriate media supplemented with plant growth regulators to induce proliferation of multiple shoots, followed by rooting of the excised shoots to regenerate whole plants,
2. The method in which totipotency of cells is realized in the form of *de novo* organogenesis, either directly in the form of induction of shoot meristems on the explants or indirectly via a callus (unorganised mass of cells resulting from proliferation of cells of the explant) and plants are regenerated through induction of roots on the resultant shoots,
3. Somatic embryogenesis, in which asexual adventive embryos (comparable to zygotic embryos in their structure and development) are induced directly on explants or indirectly through a callus phase.

The first method involving the meristems and induction of multiple shoots is the preferred method for the micropropagation industry since the risks of somaclonal variation (genetic variation induced in tissue culture) are minimal when compared to the other two methods. Somatic embryogenesis is a method that has the potential to be several times higher in multiplication rates and is amenable to handling in liquid culture systems like bioreactors.

Some explants, like the root tip, are hard to isolate and are contaminated with soil microflora that become problematic during the tissue culture process. Certain soil microflora can form tight associations with the root systems, or even grow within the root. Soil particles bound to roots are difficult to remove without injury to the roots that then allows microbial attack. These associated microflora will generally overgrow the tissue culture medium before there is significant growth of plant tissue.

Aerial (above soil) explants are also rich in undesirable microflora. However, they are more easily removed from the explant by gentle rinsing, and the remainder usually can be killed by surface sterilization. Most of the surface microflora do not form tight associations with the plant tissue. Such associations can usually be found by visual inspection as a mosaic, de-colorization or localized necrosis on the surface of the explant.

An alternative for obtaining uncontaminated explants is to take explants from seedlings which are aseptically grown from surface-sterilized seeds. The hard surface of the seed is less permeable to penetration of harsh surface sterilizing agents, such as hypochlorite, so the acceptable conditions of sterilization used for seeds can be much more stringent than for vegetative tissues.

Tissue cultured plants are clones. If the original mother plant used to produce the first explants is susceptible to a pathogen or environmental condition, the entire crop would be susceptible to the same problem. Conversely, any positive traits would remain within the line also.

Applications

Plant tissue culture is used widely in the plant sciences, forestry, and in horticulture. Applications include:

- The commercial production of plants used as potting, landscape, and florist subjects, which uses meristem and shoot culture to produce large numbers of identical individuals.

- To conserve rare or endangered plant species.
- A plant breeder may use tissue culture to screen cells rather than plants for advantageous characters, e.g. herbicide resistance/tolerance.
- Large-scale growth of plant cells in liquid culture in bioreactors for production of valuable compounds, like plant-derived secondary metabolites and recombinant proteins used as biopharmaceuticals.
- To cross distantly related species by protoplast fusion and regeneration of the novel hybrid.
- To cross-pollinate distantly related species and then tissue culture the resulting embryo which would otherwise normally die (Embryo Rescue).
- For production of doubled monoploid (dihaploid) plants from haploid cultures to achieve homozygous lines more rapidly in breeding programmes, usually by treatment with colchicine which causes doubling of the chromosome number.
- As a tissue for transformation, followed by either short-term testing of genetic constructs or regeneration of transgenic plants.
- Certain techniques such as meristem tip culture can be used to produce clean plant material from virused stock, such as potatoes and many species of soft fruit.
- Production of identical sterile hybrid species can be obtained.

Laboratories

Although some growers and nurseries have their own labs for propagating plants by the technique of tissue culture, a number of independent laboratories provide custom propagation services. The Plant Tissue Culture Information Exchange lists many commercial tissue culture labs. Since plant tissue culture is a very labour intensive process, this would be an important factor in determining which plants would be commercially viable to propagate in a laboratory.

Microorganism

A microorganism or microbe is a microscopic organism, which may be a single cell or multicellular organism. The study of microorganisms is called microbiology, a subject that began with Antonie van Leeuwenhoek's discovery of microorganisms in 1675, using a microscope of his own design.

Microorganisms are very diverse; they include all of the prokaryotes, namely the bacteria and archaea; and various forms of eukaryotes, comprising the protozoa, fungi, algae, microscopic plants (green algae), and animals such as rotifers and planarians. Some microbiologists also classify viruses as microorganisms, but others consider these as nonliving. Most microorganisms are microscopic, but there are some like *Thiomargarita namibiensis*, which are macroscopic and visible to the naked eye.

Microorganisms live in every part of the biosphere including soil, hot springs, on the ocean floor, high in the atmosphere and deep inside rocks within the Earth's crust. Microorganisms are crucial to nutrient recycling in ecosystems as they act as decomposers. As some microorganisms can fix nitrogen, they are a vital part of the nitrogen cycle, and recent studies indicate that airborne microbes may play a role in precipitation and weather.

On 17 March 2013, researchers reported data that suggested microbial life forms thrive in the Mariana Trench. the deepest spot in the Earth's oceans. Other researchers reported related studies that microbes thrive inside rocks up to 1900 feet (580 metres) below the sea floor under 8500 feet (2590 metres) of ocean off the coast of the northwestern United States. According to one of the researchers,"You can find microbes everywhere — they're extremely adaptable to conditions, and survive wherever they are."

Microbes are also exploited by people in biotechnology, both in traditional food and beverage preparation, and in modern technologies based on genetic engineering. However there are many pathogenic microbes which are harmful and can even cause death in plants and animals.

History

Evolution: Single-celled microorganisms were the first forms of life to develop on Earth, approximately 3–4 billion years ago. Further evolution was slow, and for about 3 billion years in the Precambrian eon, all organisms were microscopic. So, for most of the history of life on Earth the only forms of life were microorganisms. Bacteria, algae and fungi have been identified in amber that is 220 million years old, which shows that the morphology of microorganisms has changed little since the Triassic period.

Microorganisms tend to have a relatively fast rate of evolution. Most microorganisms can reproduce rapidly, and bacteria are also able to freely exchange genes through conjugation, transformation and transduction, even between widely divergent species. This horizontal gene transfer, coupled with a high mutation rate and many other means of genetic variation, allows microorganisms to swiftly evolve (via natural selection) to survive in new environments and respond to environmental stresses. This rapid evolution is important in medicine, as it has led to the recent development of "super-bugs", pathogenic bacteria that are resistant to modern antibiotics.

Pre-microbiology

The possibility that microorganisms exist was discussed for many centuries before their actual discovery in the 17th century. The existence of unseen microbiological life was postulated by Jainism, which is based on Mahavira's teachings as early as 6th century BCE. Paul Dundas notes that Mahavira asserted existence of unseen microbiological creatures living in earth, water, air and fire. Jain scriptures also describe nigodas, which are sub-microscopic creatures living in large clusters and having a very short life and are said to pervade each and every part of universe, even in tissues of plants and flesh of animals. However, the earliest known idea to indicate the possibility of diseases spreading by yet unseen organisms was that of the Roman scholar Marcus Terentius Varro in a 1st-century BC book titled *On Agriculture* in which he warns against locating a homestead near swamps:

... and because there are bred certain minute creatures that cannot be seen by the eyes, which float in the air and enter the body through the mouth and nose and they cause serious diseases.

In *The Canon of Medicine* (1020), Abû Alî ibn Sînâ (Avicenna) hypothesized that tuberculosis and other diseases might be contagious

In 1546, Girolamo Fracastoro proposed that epidemic diseases were caused by transferable seedlike entities that could transmit infection by direct or indirect contact, or even without contact over long distances.

All these early claims about the existence of microorganisms were speculative and were not based on any data or science. Microorganisms were neither proven, observed, nor correctly and accurately described until the 17th century. The reason for this was that all these early studies lacked the microscope.

History of Microorganisms' Discovery

Antonie Van Leeuwenhoek (1632–1723) was one of the first people to observe microorganisms, using microscopes of his own design.

Robert Hooke, a contemporary of Leeuwenhoek, also used microscopes to observe microbial life; his 1665 book *Micrographia* describes these observations and coined the term *cell*. Before Leeuwenhoek's discovery of microorganisms in 1675, it had been a mystery why grapes could be turned into wine, milk into cheese, or why food would spoil. Leeuwenhoek did not make the connection between these processes and microorganisms, but using a microscope, he did establish that there were forms of life that were not visible to the naked eye. Leeuwenhoek's discovery, along with subsequent observations by Spallanzani and Pasteur, ended the long-held belief that life spontaneously appeared from non-living substances during the process of spoilage.

Lazzaro Spallanzani (1729–1799) found that boiling broth would sterilise it, killing any microorganisms in it. He also found that new microorganisms could only settle in a broth if the broth was exposed to air. Louis Pasteur (1822–1895) expanded upon Spallanzani's findings by exposing boiled broths to the air, in vessels that contained a filter to prevent all particles from passing through to the growth medium, and also in vessels with no filter at all, with air being admitted via a curved tube that would not allow dust particles to come in contact with the broth. By boiling the broth beforehand, Pasteur ensured that no microorganisms survived within the broths at the beginning of his experiment. Nothing grew in the broths in the course of Pasteur's experiment. This meant that the living organisms that grew in such broths came from outside, as spores on dust, rather than spontaneously generated within the broth. Thus, Pasteur dealt the death blow to the theory of spontaneous generation and supported germ theory.

In 1876, Robert Koch (1843–1910) established that microbes can cause disease. He found that the blood of cattle who were infected with anthrax always had large numbers of *Bacillus anthracis*. Koch found that he could transmit anthrax from one animal to another by taking a small sample of blood from the infected animal and injecting it into a healthy one, and this caused the healthy animal to become sick. He also found that he could grow the bacteria in a nutrient broth, then inject it into a healthy animal, and cause illness. Based on these experiments, he devised criteria for establishing a causal link between a microbe and a disease and these are now known as Koch's postulates. Although these postulates cannot be applied in all cases, they do retain historical importance to the development of scientific thought and are still being used today.

4

Classification and Structure

Microorganisms can be found almost anywhere in the taxonomic organization of life on the planet. Bacteria and archaea are almost always microscopic, while a number of eukaryotes are also microscopic, including most protists, some fungi, as well as some animals and plants. Viruses are generally regarded as not living and therefore not considered as microbes, although the field of microbiology also encompasses the study of viruses.

Prokaryotes

Prokaryotes are organisms that lack a cell nucleus and the other membrane bound organelles. They are almost always unicellular, although some species such as myxobacteria can aggregate into complex structures as part of their life cycle.

Consisting of two domains, bacteria and archaea, the prokaryotes are the most diverse and abundant group of organisms on Earth and inhabit practically all environments where the temperature is below +140 °C. They are found in water, soil, air, animals' gastrointestinal tracts, hot springs and even deep beneath the Earth's crust in rocks. Practically all surfaces that have not been specially sterilized are covered by prokaryotes. The number of prokaryotes on Earth is estimated to be around five million trillion trillion, or 5×10^{30}, accounting for at least half the biomass on Earth.

Bacteria

Almost all bacteria are invisible to the naked eye, with a few extremely rare exceptions, such as *Thiomargarita namibiensis*. They lack a nucleus and other membrane-bound organelles, and can function and reproduce as individual cells, but often aggregate in multicellular colonies. Their genome is usually a single loop of DNA, although they can also harbor small pieces of DNA called plasmids. These plasmids can be transferred between cells through bacterial conjugation. Bacteria are surrounded by a cell wall, which provides strength and rigidity to their cells. They reproduce by binary fission or sometimes by budding, but do not undergo meiotic sexual reproduction. However, many bacterial species can transfer DNA between individual cells by a process referred to as natural transformation. In nature, the development of competence for transformation is usually associated with stressful environmental conditions, and seems to be an adaptation for facilitating repair of DNA damage in recipient cells. Some species form extraordinarily resilient spores, but for bacteria this is a mechanism for survival, not reproduction. Under optimal conditions bacteria can grow extremely rapidly and can double as quickly as every 20 minutes.

Archaea

Archaea are also single-celled organisms that lack nuclei. In the past, the differences between bacteria and archaea were not recognised and archaea were classified with bacteria as part of the kingdom Monera. However, in 1990 the microbiologist Carl Woese proposed the three-domain system that divided living things into bacteria, archaea and eukaryotes. Archaea differ from bacteria in both their genetics and biochemistry. For example, while bacterial cell membranes are made from phosphoglycerides with ester bonds, archaean membranes are made of ether lipids.

Archaea were originally described in extreme environments, such as hot springs, but have since been found in all types of habitats. Only now are scientists beginning to realize how common archaea are in the environment, with crenarchaeota being the most common form of life in the ocean, dominating ecosystems below 150 m in depth. These organisms are also common in soil and play a vital role in ammonia oxidation.

Eukaryotes

Most living things that are visible to the naked eye in their adult form are eukaryotes, including humans. However, a large number of eukaryotes are also microorganisms. Unlike bacteria and archaea, eukaryotes contain organelles such as the cell nucleus, the Golgi apparatus and mitochondria in their cells. The nucleus is an organelle that houses the DNA that makes up a cell's genome. DNA itself is arranged in complex chromosomes. Mitochondria are organelles vital in metabolism as they are the site of the citric acid cycle and oxidative phosphorylation. They evolved from symbiotic bacteria and retain a remnant genome. Like bacteria, plant cells have cell walls, and contain organelles such as chloroplasts in addition to the organelles in other eukaryotes. Chloroplasts produce energy from light by photosynthesis, and were also originally symbiotic bacteria.

Unicellular eukaryotes consist of a single cell throughout their life cycle. This qualification is significant since most multicellular eukaryotes consist of a single cell called a zygote only at the beginning of their life cycles. Microbial eukaryotes can be either haploid or diploid, and some organisms have multiple cell nuclei.

Unicellular eukaryotes usually reproduce asexually by mitosis under favourable conditions. However, under stressful conditions such as nutrient limitations and other conditions associated with DNA damage, they tend to reproduce sexually by meiosis and syngamy.

Protists

Of eukaryotic groups, the protists are most commonly unicellular and microscopic. This is a highly diverse group of organisms that are not easy to classify. Several algae species are multicellular protists, and slime molds have unique life cycles that involve switching between unicellular, colonial, and multicellular forms. The number of species of protists is unknown since we may have identified only a small proportion. Studies from 2001-2004 have shown that a large number of protist diversity exitsts in oceans, deep sea-vents, river sediment and an acidic river which suggests that a large number of eukaryotic microbial communities have yet to be discovered.

Animals

Most animals are multicellular, but some are too small to be seen by the naked eye. Microscopic arthropods include dust mites and spider mites. Microscopic crustaceans include copepods and the cladocera, while many nematodes are too small to be seen with the naked eye. Another particularly common group of microscopic animals are the rotifers, which are filter feeders that are usually found in fresh water. Micro-animals reproduce both sexually and asexually and may reach new habitats as some eggs can survive harsh environments that would kill the adult animal. However, some simple animals, such as rotifers and nematodes, can dry out completely and remain dormant for long periods of time.

Fungi

The fungi have several unicellular species, such as baker's yeast (*Saccharomyces cerevisiae*) and fission yeast (*Schizosaccharomyces pombe*). Some fungi, such as the pathogenic yeast *Candida albicans*, can undergo phenotypic switching and grow as single cells in some environments, and filamentous hyphae in others. Fungi reproduce both asexually, by budding or binary fission, as well by producing spores, which are called conidia when produced asexually, or basidiospores when produced sexually.

Plants

The green algae are a large group of photosynthetic eukaryotes that include many microscopic organisms. Although some green algae are classified as protists, others such as charophyta are classified with embryophyte plants, which are the most familiar group of land plants. Algae can grow as single cells, or in long chains of cells. The green algae include unicellular and colonial flagellates, usually but not always with two flagella per cell, as well as various colonial, coccoid, and filamentous forms. In the Charales, which are the algae most closely related to higher plants, cells differentiate into several distinct tissues within the organism. There are about 6000 species of green algae.

Habitats and Ecology

Microorganisms are found in almost every habitat present in nature. Even in hostile environments such as the poles, deserts, geysers, rocks, and the deep sea. Some types of microorganisms have adapted to the extreme conditions and sustained colonies; these organisms are known as extremophiles. Extremophiles have been isolated from rocks as much as 7 kilometres below the Earth's surface, and it has been suggested that the amount of living organisms below the Earth's surface may be comparable with the amount of life on or above the surface.

Extremophiles have been known to survive for a prolonged time in a vacuum, and can be highly resistant to radiation, which may even allow them to survive in space. Many types of microorganisms have intimate symbiotic relationships with other larger organisms; some of which are mutually beneficial (mutualism), while others can be damaging to the host organism (parasitism). If microorganisms can cause disease in a host they are known as pathogens and then they are usually referred to as microbes.

Extremophiles

Extremophiles are microorganisms that have adapted so that they can survive and even thrive in conditions that are normally fatal to most life-forms.

For example, some species have been found in the following extreme environments:

- Temperature: as high as 130 °C (266 °F), as low as –17 °C (1 °F)
- Acidity/alkalinity: less than pH 0, up to pH 11.5
- Salinity: up to saturation
- Pressure: up to 1,000-2,000 atm, down to 0 atm (e.g. vacuum of space)
- Radiation: up to 5kGy

Extremophiles are significant in different ways. They extend terrestrial life into much of the Earth's hydrosphere, crust and atmosphere, their specific evolutionary adaptation mechanisms to their extreme environment can be exploited in bio-technology, and their very existence under such extreme conditions increases the potential for extraterrestrial life.

Soil Microorganisms

The nitrogen cycle in soils depends on the fixation of atmospheric nitrogen. One way this can occur is in the nodules in the roots of legumes that contain symbiotic bacteria of the genera *Rhizobium*, *Mesorhizobium*, *Sinorhizobium*, *Bradyrhizobium*, and *Azorhizobium*.

Symbiotic Microorganisms

Symbiotic microbes such as fungi and algae form an association in lichen. Certain fungi form mycorrhizal symbioses with trees that increase the supply of nutrients to the tree.

Importance

Microorganisms are vital to humans and the environment, as they participate in the Earth's element cycles such as the carbon cycle and nitrogen cycle, as well as fulfilling other vital roles in virtually all ecosystems, such as recycling other organisms' dead remains and waste products through decomposition. Microorganisms also have an important place in most higher-order multicellular organisms as symbionts. Many blame the failure of Biosphere 2 on an improper balance of microorganisms.

Use in Digestion

Some forms of bacteria that live in animals' stomachs help in their digestion. For example,cows have a variety of different microbes in their stomachs that aid them in their digestion of grass and hay.

Use in Food

Microorganisms are used in brewing, winemaking, baking, pickling and other food-making processes.

They are also used to control the fermentation process in the production of cultured dairy products such as yogurt and cheese. The cultures also provide flavour and aroma, and inhibit undesirable organisms.

Use in Water Treatment

The majority of all oxidative sewage treatment processes rely on a large range of microorganisms to oxidise organic constituents which are not amenable to sedimentation or flotation. Anaerobic microorganisms are also used to reduce sludge solids producing methane gas (amongst other gases) and a sterile mineralised residue. In potable water treatment, one method, the slow sand filter, employs a complex gelatinous layer composed of a wide range of microorganisms to remove both dissolved and particulate material from raw water.

Use in Energy

Microbes are used in fermentation to produce ethanol, and in biogas reactors to produce methane. Scientists are researching the use of algae to produce liquid fuels, and bacteria to convert various forms of agricultural and urban waste into usable fuels.

Use in Production of Chemicals, Enzymes etc.

Many microbes are used for commercial and industrial production of chemicals, enzymes and other bioactive molecules.

Examples of organic acid produced include

- Acetic acid: Produced by the bacterium *Acetobacter aceti* and other acetic acid bacteria (AAB)
- Butyric acid (butanoic acid): Produced by the bacterium *Clostridium butyricum*
- Lactic acid: *Lactobacillus* and others commonly called as lactic acid bacteria (LAB)
- Citric acid: Produced by the fungus *Aspergillus niger*

Microbes are used for preparation of bioactive molecules and enzymes.

- Streptokinase produced by the bacterium *Streptococcus* and modified by genetic engineering is used as a clot buster for removing clots from the blood vessels of patients who have undergone myocardial infarctions leading to heart attack.
- Cyclosporin A is a bioactive molecule used as an immunosuppressive agent in organ transplantation
- Statins produced by the yeast *Monascus purpureus* are commercialised as blood cholesterol lowering agents which act by competitively inhibiting the enzyme responsible for synthesis of cholesterol.

Use in Science

Microbes are also essential tools in biotechnology, biochemistry, genetics, and molecular biology. The yeasts (*Saccharomyces cerevisiae*) and fission yeast (*Schizosaccharomyces pombe*) are important model organisms in science, since they are simple eukaryotes that can be grown rapidly in large numbers and are easily manipulated. They are particularly valuable in genetics, genomics and proteomics. Microbes can be harnessed for uses such as creating steroids and treating skin diseases. Scientists are also considering using microbes for living fuel cells, and as a solution for pollution.

Use in Warfare

In the Middle Ages, diseased corpses were thrown into castles during sieges using catapults or other siege engines. Individuals near the corpses were exposed to the deadly pathogen and were likely to spread that pathogen to others.

Importance in Human Health

Human Digestion

Microorganisms can form an endosymbiotic relationship with other, larger organisms. For example, the bacteria that live within the human digestive system contribute to gut immunity, synthesise vitamins such as folic acid and biotin, and ferment complex indigestible carbohydrates.

Diseases Caused by Microbes

Microorganisms are the cause of many infectious diseases. The organisms involved include pathogenic bacteria, causing diseases such as plague, tuberculosis and anthrax; protozoa, causing diseases such as malaria, sleeping sickness and toxoplasmosis; and also fungi causing diseases such as ringworm, candidiasis or histoplasmosis. However, other diseases such as influenza, yellow fever or AIDS are caused by pathogenic viruses, which are not usually classified as living organisms and are not, therefore, microorganisms by the strict definition. As of 2007, no clear examples of archaean pathogens are known, although a relationship has been proposed between the presence of some archaean methanogens and human periodontal disease.

Importance in Ecology

Microbes are critical to the processes of decomposition required to cycle nitrogen and other elements back to the natural world.

Hygiene

Hygiene is the avoidance of infection or food spoiling by eliminating microorganisms from the surroundings. As microorganisms, in particular bacteria, are found virtually everywhere, the levels of harmful microorganisms can be reduced to acceptable levels. However, in some cases, it is required that an object or substance be completely sterile, i.e. devoid of all living entities and viruses. A good example of this is a hypodermic needle.

In food preparation microorganisms are reduced by preservation methods (such as the addition of vinegar), clean utensils used in preparation, short storage periods, or by cool temperatures. If complete sterility is needed, the two most common methods are irradiation and the use of an autoclave, which resembles a pressure cooker.

There are several methods for investigating the level of hygiene in a sample of food, drinking water, equipment, etc. Water samples can be filtrated through an extremely fine filter. This filter is then placed in a nutrient medium. Microorganisms on the filter then grow to form a visible colony. Harmful microorganisms can be detected in food by placing a sample in a nutrient broth designed to enrich the organisms in question. Various methods, such as selective media or polymerase chain reaction, can then be used for detection. The hygiene of hard surfaces, such as cooking pots, can be tested by touching them with a solid piece of nutrient medium and then allowing the microorganisms to grow on it.

There are no conditions where all microorganisms would grow, and therefore often several different methods are needed. For example, a food sample might be analyzed on three different nutrient mediums designed to indicate the presence of "total" bacteria (conditions where many, but not all, bacteria grow), molds (conditions where the growth of bacteria is prevented by, e.g., antibiotics) and coliform bacteria (these indicate a sewage contamination).

Cilium

A cilium (Latin for eyelash; the plural is *cilia*) is an organelle found in eukaryotic cells. Cilia are slender protuberances that project from the much larger cell body.

There are two types of cilia: *motile* cilia and *non-motile*, or primary cilia, which typically serve as sensory organelles. In eukaryotes, motile cilia and flagella together make up a group of organelles known as undulipodia. Eukaryotic cilia are structurally identical to Eukaryotic flagella, although distinctions are sometimes made according to function and/or length.

Types and Distribution

Cilia are rare in most plants, occurring most notably in cycads.

Cilia can be divided into primary and motile forms.

Motile Cilia

Larger eukaryotes, such as mammals, have *motile* cilia as well. Motile cilia are usually present on a cell's surface in large numbers and beat in coordinated waves.

- In humans, for example, motile cilia are found in the lining of the trachea (windpipe), where they sweep mucus and dirt out of the lungs.
- In female mammals, the beating of cilia in the Fallopian tubes moves the ovum from the ovary to the uterus.

Ciliates are microscopic organisms that possess *motile* cilia exclusively and use them for either locomotion or to simply move liquid over their surface.

Primary/Immotile Cilia

In humans, *primary* cilia are found on nearly every cell in the body.

In comparison to *motile* cilia, non-motile (or *primary*) cilia usually occur one per cell; nearly all mammalian cells have a single non-motile *primary cilium*. In addition, examples of specialized primary cilia can be found in human sensory organs such as the eye and the nose:

- The outer segment of the rod photoreceptor cell in the human eye is connected to its cell body with a specialized non-motile cilium.
- The dendritic knob of the olfactory neuron, where the odorant receptors are located, also contains non-motile cilia (about 10 cilia per dendritic knob).

Although the primary cilium was discovered in 1898, it was largely ignored for a century. Only recently has great progress been made in understanding the function of the primary cilium. Until the 1990s, the prevailing view of the primary cilium was that it was merely a vestigial organelle without important function. Recent findings regarding its physiological roles in chemical sensation, signal transduction, and control of cell growth, have led scientists to acknowledge its importance in cell function, with the discovery of its role in diseases not previously recognized to involve the dysgenesis and dysfunction of cilia, such as polycystic kidney disease congenital heart disease, and an emerging group of genetic ciliopathies. The primary cilium is now known to play an important role in the function of many human organs. The current scientific understanding of primary cilia views them as "sensory cellular antennae that coordinate a large number of cellular signalling pathways, sometimes coupling the signalling to ciliary motility or alternatively to cell division and differentiation."..

Structure, Assembly, Maintenance, and Function

Inside cilia and flagella is a microtubule-based cytoskeleton called the axoneme. The axoneme of primary cilia typically has a ring of nine outer microtubule doublets (called a 9+0 axoneme), and the axoneme of a motile cilium has two central microtubule singlets in addition to the nine outer doublets (called a 9+2 axoneme). The axonemal cytoskeleton acts as a scaffolding for various protein complexes and provides binding sites for molecular motor proteins such as kinesin II, that help carry proteins up and down the microtubules.

Cilia are formed through the process of ciliogenesis. The building blocks of the cilia such as tubulins and other partially assembled axonemal proteins are added to the ciliary tips which point away from the cell body. In most species bi-directional motility called intraflagellar transport (IFT) plays an essential role to move these building materials from the cell body to the assembly site. IFT also carries the disassembled material to be recycled from the ciliary tip back to the cell body. By regulating the equilibrium between these two IFT processes, the length of cilia can be maintained dynamically. The disassembly of the cilia requires the action of the protein kinase Aurora A .

Exceptions where IFT is not present include *Plasmodium falciparum* which is one of the species of *Plasmodium* that cause malaria in humans. In this parasite, cilia assemble in the cytoplasm.

At the base of the cilium where the cilia attaches to the cell body is the microtubule organizing centre, the basal body. Some basal body proteins as CEP164, ODF2 and CEP170, regulate the formation and the stability of the cilium. A transition zone between the basal body and the axoneme "serves as a docking station for intraflagellar transport and motor proteins."

In effect, the [cilium] is a nanomachine composed of perhaps over 600 proteins in molecular complexes, many of which also function independently as nanomachines.

Ciliary Rootlet

The ciliary rootlet is a cytoskeleton-like structure that originates from the basal body at the proximal end of a cilium. It extends proximally toward the cell nucleus. Rootlets are typically 80-100 nm in diameter and contain cross striae distributed at regular intervals of approximately 55-70 nm. According to the Gene Ontology, the following proteins localize to the ciliary rootlet: amyloid precursor protein, rootletin, kinesins (KIF5B, KIF5C, KLC2, KLC3), and presenilins (PSEN1, PSEN2).

Sensing the Extracellular Environment

Some epithelial cells are ciliated, and they commonly exist as a sheet of polarized cells forming a tube or tubule with cilia projecting into the lumen." Primary cilia on epithelial cells provide chemosensation, thermosensation and mechanosensation of the extracellular environment by playing "a sensory role mediating specific signalling cues, including soluble factors in the external cell environment, a secretory role in which a soluble protein is released to have an effect downstream of the fluid flow, and mediation of fluid flow if the cilia are motile.

Cilium-related Disease

Ciliary defects can lead to a number of human diseases. Genetic mutations compromising the proper functioning of cilia, ciliopathies, can cause chronic disorders such as primary ciliary dyskinesia (PCD), nephronophthisis or Senior-Loken syndrome. In addition, a defect of the primary cilium in the renal tube cells can lead to polycystic kidney disease (PKD). In another genetic disorder called Bardet-Biedl syndrome (BBS), the mutant gene products are the components in the basal body and cilia.

Lack of functional cilia in female Fallopian tubes can cause ectopic pregnancy. A fertilized ovum may not reach the uterus if the cilia are unable to move it there. In such a case, the ovum will implant in the Fallopian tubes, causing a tubal pregnancy, the most common form of ectopic pregnancy.

Since the flagellum of human sperm is actually a modified cilium, ciliary dysfunction can also be responsible for male infertility.

Of interest, there is an association of primary ciliary dyskinesia with left-right anatomic abnormalities such as situs inversus (a combination of findings known as Kartagener's syndrome) and other heterotaxic defects. These left-right anatomic abnormalities can also result in congenital heart disease. In fact, it has been shown that proper cilial function is responsible for the normal left-right asymmetry in mammals.

Ciliopathy as an Origin for Many Multi-symptom Genetic Diseases

Recent findings in genetic research have suggested that a large number of genetic disorders, both genetic syndromes and genetic diseases, that were not previously related in the medical literature, may be, in fact, highly related in the root cause of the widely varying set of medical symptoms that are clinically visible in the disorder. These have been grouped as an emerging class of diseases called ciliopathies. The underlying cause may be a dysfunctional molecular mechanism in the primary cilia structures, organelles which are present in many diverse cellular types throughout the human body. Cilia defects adversely affect "numerous critical developmental signalling pathways" essential to cellular development and thus offer a plausible hypothesis for the often multi-symptom nature of a large set of syndromes and diseases. Known ciliopathies include primary ciliary dyskinesia, Bardet-Biedl syndrome, polycystic kidney and liver disease, nephronophthisis, Alstrom syndrome, Meckel-Gruber syndrome, Sensenbrenner syndrome and some forms of retinal degeneration.

Flagellum

A flagellum is a lash-like appendage that protrudes from the cell body of certain prokaryotic and eukaryotic cells. The word *flagellum* in Latin means whip. The primary role of the flagellum is locomotion but it also often has function as a sensory organelle, being sensitive to chemicals and temperatures outside the cell. Flagella are organelles defined by function rather than structure. There are large differences between different types of flagellum; the prokaryotic and eukaryotic flagella differ greatly in protein composition, structure, and mechanism of propulsion, however both are used for swimming.

An example of a flagellate bacterium is the ulcer-causing *Helicobacter pylori*, which uses multiple flagella to propel itself through the mucus lining to reach the stomach epithelium. An example of a eukaryotic flagellate cell is the mammalian sperm cell, which uses its flagellum to propel itself through the female reproductive tract. Eukaryotic flagella are structurally identical to eukaryotic cilia, although distinctions are sometimes made according to function and/ or length.

Types

Three types of flagella have so far been distinguished; bacterial, archaeal and eukaryotic.

The main differences among these three types are summarized below:

- Bacterial flagella are helical filaments, each with a rotary motor at its base which can turn clockwise or counterclockwise. They provide two of several kinds of bacterial motility.
- Archaeal flagella (Archaella) are superficially similar to bacterial flagella, but are different in many details and considered non-homologous.
- Eukaryotic flagella - those of animal, plant, and protist cells - are complex cellular projections that lash back and forth. Eukaryotic flagella are classed along with eukaryotic *motile* cilia as *undulipodia* to emphasize their distinctive wavy appendage role in cellular function or motility. *Primary* cilia are immotile, and are not undulipodia; they have a structurally different *9+0 axoneme* rather than the *9+2 axoneme* found in both flagella and motile cilia undulipodia.

Bacterial

Structure and composition. The bacterial flagellum is made up of the protein flagellin. Its shape is a 20 nanometer-thick hollow tube. It is helical and has a sharp bend just outside the outer membrane; this "hook" allows the axis of the helix to point directly away from the cell. A shaft runs between the hook and the basal body, passing through protein rings in the cell's membrane that act as bearings. Gram-positive organisms have 2 of these basal body rings, one in the peptidoglycan layer and one in the plasma membrane. Gram-negative organisms have 4 such rings: the L ring associates with the lipopolysaccharides, the P ring associates with peptidoglycan layer, the M ring is embedded in the plasma membrane, and the S ring is directly attached to the plasma membrane. The filament ends with a capping protein.

The flagellar filament is the long helical screw that propels the bacterium when rotated by the motor, through the hook. In most bacteria that have been studied, including the Gram negative *Escherichia coli, Salmonella typhimurium, Caulobacter crescentus,* and *Vibrio alginolyticus,* the filament is made up of eleven protofilaments approximately parallel to the filament axis. Each protofilament is a series of tandem protein chains. However in *Campylobacter jejuni,* there are seven protofilaments.

The basal body has several traits in common with some types of secretory pores, such as the hollow rod-like "plug" in their centres extending out through the plasma membrane. Given the structural similarities between bacterial flagella and bacterial secretory systems, it is thought that bacterial flagella may have evolved from the type three secretion system; however, it is not known for certain whether these pores are derived from the bacterial flagella or the bacterial secretory system.

Motor: The bacterial flagellum is driven by a rotary engine (the *Mot complex*) made up of protein, located at the flagellum's anchor point on the inner cell membrane. The engine is powered by proton motive force, i.e., by the flow of protons (hydrogen ions) across the bacterial cell membrane due to a concentration gradient set up by the cell's metabolism (in *Vibrio* species there are two kinds of flagella, lateral and polar, and some are driven by a sodium ion pump rather than a proton pump). The rotor transports protons across the membrane, and is turned in the process. The rotor alone can operate at 6,000 to 17,000 rpm, but with the flagellar filament attached usually only reaches 200 to 1000 rpm. The direction of rotation can be switched almost instantaneously, caused by a slight change in the position of a protein, FliG, in the rotor.

The cylindrical shape of flagella is suited to locomotion of microscopic organisms; these organisms operate at a low Reynolds number, where the viscosity of the surrounding water is much more important than its mass or inertia.

The rotational speed of flagella varies in response to the intensity of the proton motive force, thereby permitting certain forms of speed control, and also permitting some types of bacteria to attain remarkable speeds in proportion to their size; some achieve roughly 60 cell lengths / second. Although at such a speed it would take a bacterium about 245 days to cover a kilometre, and although that may seem slow, the perspective changes when the concept of scale is introduced. In comparison to macroscopic life forms it is very fast indeed when expressed in terms of number of body lengths per second. A cheetah for example, only achieves about 25 body lengths / sec.

Through use of their flagella, *E. coli* are able to move rapidly towards attractants and away from repellents. They do this by means of a biased random walk, with 'runs' and 'tumbles' brought about by rotating the flagellum counterclockwise and clockwise respectively.

Assembly: During flagellar assembly, components of the flagellum pass through the hollow cores of the basal body and the nascent filament. During assembly, protein components are added at the flagellar tip rather than at the base. *In vitro*, flagellar filaments assemble spontaneously in a solution containing purified flagellin as the sole protein.

Evolution: The evolution of bacterial flagella has been used as an argument against evolution by creationists. They argue that complex structures like flagella cannot evolve from simple structures. In other words, flagella are "irreducibly complex" and need all of their protein components to function. However, it has been shown by numerous studies that a large number of proteins can be deleted without (complete) loss of function. Moreover, it is generally accepted now that bacterial flagella have evolved from much simpler secretion systems, such as the Type III secretion system.

Flagella Arrangement Schemes

Different species of bacteria have different numbers and arrangements of flagella. Monotrichous bacteria have a single flagellum (e.g., *Vibrio cholerae*). Lophotrichous bacteria have multiple flagella located at the same spot on the bacteria's surfaces which act in concert to drive the bacteria in a single direction. In many cases, the bases of multiple flagella are surrounded by a specialized region of the cell membrane, the so-called *polar organelle*. Amphitrichous bacteria have a single flagellum on each of two opposite ends (only one flagellum operates at a time, allowing the bacteria to reverse course rapidly by switching which flagellum is active). Peritrichous bacteria have flagella projecting in all directions (e.g., *E. coli*).

In certain large forms of *Selenomonas*, more than 30 individual flagella are organized outside the cell body, helically twining about each other to form a thick structure (easily visible with the light microscope) called a "fascicle". Other bacteria, such as most Spirochetes, have two or more specialized flagella (endoflagella) arising from opposite poles of the cell, which together constitute the so-called "axial filament" that is located within the periplasmic space between the flexible cell wall and an outer sheath. The rotation of the axial filament relative to the cell body causes the entire bacterium to move forward in a corkscrew-like motion, even through material viscous enough to prevent the passage of normally flagellated bacteria.

Counterclockwise rotation of a monotrichous polar flagellum pushes the cell forward with the flagellum trailing behind, much like a corkscrew moving inside cork. Indeed water on the microscopic scale is highly viscous, very different from our daily experience of water.

Flagella are left-handed helices, and bundle and rotate together only when rotating counterclockwise. When some of the rotors reverse direction, the flagella unwind and the cell starts "tumbling". It has also been suggested that even if all flagella would rotate clockwise, they will not form a bundle, due to geometrical as well as hydrodynamic reasons. Such "tumbling" may happen occasionally, leading to the cell seemingly thrashing about in place, resulting in the reorientation of the cell. The clockwise rotation of a flagellum is suppressed by chemical compounds favourable to the cell (e.g. food), but the motor is highly adaptive to this. Therefore, when moving in a favourable direction, the concentration of the chemical attractant increases and "tumbles" are continually suppressed; however, when the cell's direction of motion is unfavourable (e.g., away from a chemical attractant), tumbles are no longer suppressed and occur much more often, with the chance that the cell will be thus reoriented in the correct direction.

In some *Vibrio* spp. (particularly *Vibrio parahemolyticus*) and related proteobacteria such as *Aeromonas*, two flagellar systems co-exist, using different sets of genes and different ion gradients for energy. The polar flagella are constitutively expressed and provide motility in bulk fluid, while the lateral flagella are expressed when the polar flagella meet too much resistance to turn. These provide swarming motility on surfaces or in viscous fluids.

Archaeal

The archaeal flagellum (Archaellum) is superficially similar to the bacterial (or eubacterial) flagellum; in the 1980s they were thought to be homologous on the basis of gross morphology and behaviour. Both flagella and archaella consist of filaments extending outside the cell, and rotate to propel the cell. Archaeal flagella have a unique structure which lacks a central channel. Similar to bacterial type IV pilins, the archaeal flagellins (archaellins) are made with class 3 signal peptides and they are processed by a type IV prepilin peptidase-like enzyme. The archaellins are typically modified by the addition of N-linked glycans which are necessary for proper assembly and/or function.

Discoveries in the 1990s revealed numerous detailed differences between the archaeal and bacterial flagella; these include:

- Bacterial flagella are motorized by a flow of H^+ ions (or occasionally Na^+ ions); archaeal flagella are almost certainly powered by ATP. The torque-generating motor that powers rotation of the archaeal flagellum has not been identified.
- While bacterial cells often have many flagellar filaments, each of which rotates independently, the archaeal flagellum is composed of a bundle of many filaments that rotate as a single assembly.
- Bacterial flagella grow by the addition of flagellin subunits at the tip; archaeal flagella grow by the addition of subunits to the base.
- Bacterial flagella are thicker than archaella [Singular Archaellum], and the bacterial filament has a large enough hollow "tube" inside that the flagellin subunits can flow up the inside of the filament and get added at the tip; the archaellum is too thin (12-15 nm) to allow this.
- Many components of bacterial flagella share sequence similarity to components of the type III secretion systems, but the components of bacterial flagella and archaella share no sequence similarity. Instead, some components of archaella share sequence and morphological similarity with components of type IV pili, which are assembled through the action of type II secretion systems (the nomenclature of pili and protein secretion systems is not consistent).

These differences could mean that the bacterial flagella and archaella could be a classic case of biological analogy, or convergent evolution, rather than homology. However, in comparison to the decades of well-publicized study of bacterial flagella (e.g. by Howard Berg), archaella have only recently begun to get serious scientific attention.

Therefore, many assume erroneously that there is only one basic kind of prokaryotic flagellum, and that archaella are homologous to it. For example, Cavalier-Smith (2002) is aware of the differences between bacterial flagellins and archaeal flagellins (archaellins).

Eukaryotic

Along with motile cilia, flagella make up a group of organelles known as undulipodia.

Structure

A eukaryotic flagellum is a bundle of nine fused pairs of microtubule *doublets* surrounding two central single microtubules. The so-called *"9+2"* structure is characteristic of the core of the eukaryotic flagellum called an *axoneme*. At the base of a eukaryotic flagellum is a basal body, "blepharoplast" or kinetosome, which is the microtubule organizing centre (MTOC) for flagellar microtubules and is about 500 nanometers long. Basal bodies are structurally identical to centrioles. The flagellum is encased within the cell's plasma membrane, so that the interior of the flagellum is accessible to the cell's cytoplasm.

Mechanism

Each of the outer 9 doublet microtubules extends a pair of dynein arms (an "inner" and an "outer" arm) to the adjacent microtubule; these dynein arms are responsible for flagellar beating, as the force produced by the arms causes the microtubule doublets to slide against each other and the flagellum as a whole to bend. These dynein arms produce force through ATP hydrolysis. The flagellar axoneme also contains radial spokes, polypeptide complexes extending from each of the outer 9 microtubule doublets towards the central pair, with the "head" of the spoke facing inwards. The radial spoke is thought to be involved in the regulation of flagellar motion, although its exact function and method of action are not yet understood.

Flagella vs Cilia

The regular beat patterns of eukaryotic cilia and flagella generate motion on a cellular level. Examples range from the propulsion of single cells such as the swimming of spermatozoa to the transport of fluid along a stationary layer of cells such as in the respiratory tract. Though eukaryotic flagella and motile cilia are ultrastructurally identical, the beating pattern of the two organelles can be different. In the case of flagella the motion is often planar and wave-like, whereas the motile cilia often perform a more complicated 3D motion with a power and recovery stroke.

Intraflagellar Transport

Intraflagellar transport (IFT), the process by which axonemal subunits, transmembrane receptors, and other proteins are moved up and down the length of the flagellum, is essential for proper functioning of the flagellum, in both motility and signal transduction.

Plant Virus

Plant viruses are viruses that affect plants. Like all other viruses, plant viruses are obligate intracellular parasites that do not have the molecular machinery to replicate without a host. Plant viruses are pathogenic to higher plants. While this article does not intend to list all plant viruses, it discusses some important viruses as well as their uses in plant molecular biology.

Overview

Although plant viruses are not nearly as well understood as the animal counterparts, one plant virus has become iconic. The first virus to be discovered was *Tobacco mosaic virus* (TMV). This and other viruses cause an estimated US$60 billion loss in crop yields worldwide each year. Plant viruses are grouped into 73 genera and 49 families.

To transmit from one plant to another and from one plant cell to another, plant viruses must use strategies that are usually different from animal viruses. Plants do not move, and so plant-to-plant transmission usually involves vectors (such as insects). Plant cells are surrounded by solid cell walls, therefore transport through plasmodesmata is the preferred path for virions to move between plant cells. Plants probably have specialized mechanisms for transporting mRNAs through plasmodesmata, and these mechanisms are thought to be used by RNA viruses to spread from one cell to another.

Plant defences against viral infection include, among other measures, the use of siRNA in response to dsRNA. Most plant viruses encode a protein to suppress this response. Plants also reduce transport through plasmodesmata in response to injury.

History

The discovery of plant viruses causing disease is often accredited to Martinus Beijerinck who determined, in 1898, that plant sap obtained from tobacco leaves with the "mosaic disease" remained infectious when passed through a porcelain filter. This was in contrast to bacteria microorganisms, which were retained by the filter. Beijerinck referred to the infectious filtrate as a "contagium vivum fluidum", thus the coinage of the modern term "virus". After the initial discovery of the 'viral concept' there was need to classify any other known viral diseases based on the mode of transmission even though microscopic observation proved fruitless. In 1939 Holmes published a classification list of 129 plant viruses. This was expanded and in 1999 there were 977 officially recognized, and some provisional, plant virus species.

The purification (crystallization) of TMV was first performed by Wendell Stanley, who published his findings in 1935, although he did not determine that the RNA was the infectious material. However, he received the Nobel Prize in Chemistry in 1946. In the 1950s a discovery by two labs simultaneously proved that the purified RNA of the TMV was infectious which reinforced the argument. The RNA carries genetic information to code for the production of new infectious particles.

More recently virus research has been focused on understanding the genetics and molecular biology of plant virus genomes, with a particular interest in determining how the virus can replicate, move and infect plants. Understanding the virus genetics and protein functions has been used to explore the potential for commercial use by biotechnology companies. In particular, viral-derived sequences have been used to provide an understanding of novel forms of resistance. The recent boom in technology allowing humans to manipulate plant viruses may provide new strategies for production of value-added proteins in plants.

Structure

Viruses are extremely small and can only be observed with an electron microscope. The structure of a virus is given by its coat of proteins, which surround the viral genome. Assembly of viral particles takes place spontaneously. Over 50% of known plant viruses are rod-shaped (flexuous or rigid). The length of the particle is normally dependent on the genome but it is usually between 300–500 nm with a diameter of 15–20 nm. Protein subunits can be placed around the circumference of a circle to form a disc. In the presence of the viral genome, the discs are stacked, then a tube is created with room for the nucleic acid genome in the middle.

The second most common structure amongst plant viruses are isometric particles. They are 40–50 nm in diameter. In cases when there is only a single coat protein, the basic structure consists of 60 T subunits, where T is an integer. Some viruses may have 2 coat proteins are the associate to form an icosahedral shaped particle. There are three genera of *Geminiviridae* that possess geminate particles which are like two isometric particles stuck together. A very small number of plant viruses have, in addition to their coat proteins, a lipid envelope. This is derived from the plant cell membrane as the virus particle buds off from the cell.

Transmission of Plant Viruses

Through Sap

Viruses can be spread by direct transfer of sap by contact of a wounded plant with a healthy one. Such contact may occur during agricultural practices, as by damage caused by tools or hands, or naturally, as by an animal feeding on the plant. Generally TMV, potato viruses and cucumber mosaic viruses are transmitted via sap.

Insects

Plant viruses need to be transmitted by a vector, most often insects such as leafhoppers. One class of viruses, the Rhabdoviridae, has been proposed to actually be insect viruses that have evolved to replicate in plants. The chosen insect vector of a plant virus will often be the determining factor in that virus's host range: it can only infect plants that the insect vector feeds upon. This was shown in part when the old world white fly made it to the USA, where it transferred many plant viruses into new hosts. Depending on the way they are transmitted, plant viruses are classified as non-persistent, semi-persistent and persistent. In non-persistent transmission, viruses become attached to the distal tip of the stylet of the insect and on the next plant it feeds on, it inoculates it with the virus. Semi-persistent viral transmission involves the virus entering the foregut of the insect.

Those viruses that manage to pass through the gut into the haemolymph and then to the salivary glands are known as persistent. There are two sub-classes of persistent viruses: propagative and circulative. Propagative viruses are able to replicate in both the plant and the insect (and may have originally been insect viruses), whereas circulative can not. Circulative viruses are protected inside aphids by the chaperone protein symbionin produced by bacterial symbionts. Many plant viruses encode within their genome polypeptides with domains essential for transmission by insects. In non-persistent and semi-persistent viruses, these domains are in the coat protein and another protein known as the helper component. A bridging hypothesis has been proposed to explain how these proteins aid in insect-mediated viral transmission. The helper component will bind to the specific domain of the coat protein, and then the insect mouthparts — creating a bridge. In persistent propagative viruses, such as tomato spotted wilt virus (TSWV), there is often a lipid coat surrounding the proteins that is not seen in the other classes of plant viruses. In the case of TSWV, 2 viral proteins are expressed in this lipid envelope. It has been proposed that the viruses bind via these proteins and are then taken into the insect cell by receptor-mediated endocytosis.

Nematodes

Soil-borne nematodes also have been shown to transmit viruses. They acquire and transmit them by feeding on infected roots. Viruses can be transmitted both non-persistently and persistently, but there is no evidence of viruses being able to replicate in nematodes. The virions attach to the stylet (feeding organ) or to the gut when they feed on an infected plant and can then unattach during later feeding to infect other plants. Examples of viruses that can be transmitted by nematodes include tobacco ringspot virus and tobacco rattle virus.

Plasmodiophorids

A number of virus genera are transmitted, both persistently and non-persistently, by soil borne zoosporic protozoa. These protozoa are not phytopathogenic themselves, but parasitic. Transmission of the virus takes place when they become associated with the plant roots. Examples include *Polymyxa graminis,* which has been shown to transmit plant viral diseases in cereal crops and *Polymyxa betae* which transmits Beet necrotic yellow vein virus. Plasmodiophorids also create wounds in the plant's root through which other viruses can enter.

Seed and Pollen Borne Viruses

Plant virus transmission from generation to generation occurs in about 20% of plant viruses. When viruses are transmitted by seeds, the seed is infected in the generative cells and the virus is maintained in the germ cells and sometimes, but less often, in the seed coat. When the growth and development of plants is delayed because of situations like unfavourable weather, there is an increase in the amount of virus infections in seeds. There does not seem to be a correlation between the location of the seed on the plant and its chances of being infected. Little is known about the mechanisms involved in the transmission of plant viruses via seeds, although it is known that it is environmentally influenced and that seed transmission occurs because of a direct invasion of the embryo via the ovule or by an indirect route with an attack on the embryo mediated by infected gametes. These processes can occur concurrently or separately depending on the host plant. It is unknown how the virus is able to directly invade and cross the embryo and boundary between the parental and progeny generations in the ovule. Many plants species can be infected through seeds including but not limited to the families Leguminosae, Solanaceae, Compositae, Rosaceae, Curcurbitaceae, Gramineae. Bean common mosaic virus is transmitted through seeds.

Direct Plant-to-human Transmission

Researchers from the University of the Mediterranean in Marseille, France have found evidence that suggest a virus common to peppers, the Pepper Mild Mottle Virus (PMMoV) may have moved on to infect humans. This is a very rare and highly unlikely event as, to enter a cell and replicate, a virus must "bind to a receptor on its surface, and a plant virus would be highly unlikely to recognize a receptor on a human cell. One possibility is that the virus does not infect human cells directly. Instead, the naked viral RNA may alter the function of the cells through a mechanism similar to RNA interference, in which the presence of certain RNA sequences can turn genes on and off," according to Virologist Robert Garry from the Tulane University in New Orleans, Louisiana.

Symptoms include being more likely to have fever, abdominal pain, and itching.

Translation of Plant Viral Proteins

75% of plant viruses have genomes that consist of single stranded RNA (ssRNA). 65% of plant viruses have +ssRNA, meaning that they are in the same sense orientation as messenger RNA but 10% have -ssRNA, meaning they must be converted to +ssRNA before they can be translated. 5% are double stranded RNA and so can be immediately translated as +ssRNA viruses. 3% require a reverse transcriptase enzyme to convert between RNA and DNA. 17% of plant viruses are ssDNA and very few are dsDNA, in contrast a quarter of animal viruses are dsDNA and three quarters of bacteriophage are dsDNA. Viruses use the plant ribosomes to produce the 4-10 proteins encoded by their genome. However, since all of the proteins are encoded on a single strand (that is, they are polycistronic) this will mean that the ribosome will either only produce one protein, as it will terminate translation at the first stop codon, or that a polyprotein will be produced. Plant viruses have had to evolve special techniques to allow the production of viral proteins by plant cells.

5' Cap

For translation to occur, eukaryotic mRNAs require a 5' Cap structure. This means that viruses must also have one. This normally consists of 7MeGpppN where N is normally adenine or guanine. The viruses encode a protein, normally a replicase, with a methyltransferase activity to allow this. Some viruses are cap-snatchers. During this process, a ^{7m}G-capped host mRNA is recruited by the viral transcriptase complex and subsequently cleaved by a virally encoded endonuclease. The resulting capped leader RNA is used to prime transcription on the viral genome. However some plant viruses do not use cap, yet translate efficiently due to cap-independent translation enhancers present in 5' and 3' untranslated regions of viral mRNA.

Readthrough

Some viruses (e.g. tobacco mosaic virus (TMV)) have RNA sequences that contain a "leaky" stop codon. In TMV 95% of the time the host ribosome will terminate the synthesis of the polypeptide at this codon but the rest of the time it continues past it. This means that 5% of the proteins produced are larger than and different from the others normally produced, which can be thought of as a crude form of transcriptional regulation. In TMV, this extra sequence of polypeptide is an RNA polymerase that replicates its genome.

Production of Sub-genomic RNAs

Some viruses use the production of subgenomic RNAs to ensure the translation of all proteins within their genomes. In this process the first protein encoded on the genome, and this the first to be translated, is a replicase. This protein will act on the rest of the genome producing negative strand sub-genomic RNAs then act upon these to form positive strand sub-genomic RNAs that are essentially mRNAs ready for translation.

Segmented Genomes

Some viral families, such as the *Bromoviridae* instead opt to have multi-partite genomes, genomes split between multiple viral particles. For infection to occur, the plant must be infected with all particles across the genome. For instance *Brome mosaic virus* has a genome split between 3 viral particles, and all 3 particles with the different RNAs are required for infection to take place.

Polyprotein Processing

This strategy is adopted by viral genera such as the Potyviridae and Tymoviridae.

The ribosome translates a single protein from the viral genome. Within the polyprotein is an enzyme (or enzymes) with proteinase function that is able to cleave the polyprotein into the various single proteins or just cleave away the protease, which can then cleave other polypeptides producing the mature proteins.

Well Understood Plant Viruses

Tobacco mosaic virus (TMV) and Cauliflower mosaic virus (CaMV) are frequently used in plant molecular biology. Of special interest is the CaMV 35S promoter, which is a very strong promoter most frequently used in plant transformations.

Heterospory

Heterospory is the production of spores of two different sizes and sexes by the sporophytes of land plants. Heterospory was evolved from isospory independently by several plant groups in the Devonian period as part of the process of evolution of the timing of sex differentiation. Heterosporic plants produce small spores called microspores which either germinate to become free-living male gametophytes or have reduced male gametophytes packaged within them, and larger spores called megaspores that either germinate into free-living female gametophytes, or which have a female gametophyte packaged within them which is retained in and nurtured by the sporophyte phase, a condition referred to as endospory.

Heterosporous species are thus usually dioicous, a condition that promotes outcrossing. Some heterosporous species produce micro- and megaspores in the same sporangium, a condition known as homoangy, while in others the micro- and megaspores are produced in separate sporangia (heterangy). These may both be borne on the same monoecious sporophyte or on different sporophytes in dioecious species.

Cytogenetics

Cytogenetics is a branch of genetics that is concerned with the study of the structure and function of the cell, especially the chromosomes. It includes routine analysis of G-banded chromosomes, other cytogenetic banding techniques, as well as molecular cytogenetics such as fluorescent *in situ* hybridization (FISH) and comparative genomic hybridization (CGH).

History

Early Years: Chromosomes were first observed in plant cells by Karl Wilhelm von Nägeli in 1842. Their behaviour in animal (salamander) cells was described by Walther Flemming, the discoverer of mitosis, in 1882. The name was coined by another German anatomist, von Waldeyer in 1888.

The next stage took place after the development of genetics in the early 20th century, when it was appreciated that the set of chromosomes (the karyotype) was the carrier of the genes. Levitsky seems to have been the first to define the karyotype as the phenotypic appearance of the somatic chromosomes, in contrast to their genic contents. Investigation into the human karyotype took many years to settle the most basic question: how many chromosomes does a normal diploid human cell contain? In 1912, Hans von Winiwarter reported 47 chromosomes in spermatogonia and 48 in oogonia, concluding an XX/XO sex determination mechanism. Painter in 1922 was not certain whether the diploid number of man was 46 or 48, at first favouring 46. He revised his opinion later from 46 to 48, and he correctly insisted on man having an XX/XY system. Considering their techniques, these results were quite remarkable.

In science books, the number of human chromosomes remained at 48 for over thirty years. New techniques were needed to correct this error. Joe Hin Tjio working in Albert Levan's lab was responsible for finding the approach:

1. Using cells in culture
2. Pre-treating cells in a hypotonic solution, which swells them and spreads the chromosomes
3. Arresting mitosis in metaphase by a solution of colchicine
4. Squashing the preparation on the slide forcing the chromosomes into a single plane
5. Cutting up a photomicrograph and arranging the result into an indisputable karyogram.

It took until 1956 until it became generally accepted that the karyotype of man included only 46 chromosomes. Rather interestingly, the great apes have 48 chromosomes. Human chromosome 2 was formed by a merger of ancestral chromosomes, reducing the number.

Applications in Biology

McClintock's Work on Maize: Barbara McClintock began her career as a maize cytogeneticist. In 1931, McClintock and Harriet

Creighton demonstrated that cytological recombination of marked chromosomes correlated with recombination of genetic traits (genes). McClintock, while at the Carnegie Institution, continued previous studies on the mechanisms of chromosome breakage and fusion in maize. She identified a particular chromosome breakage event that always occurred at the same locus on maize chromosome 9, which she named the "*Ds*" or "dissociation" locus. McClintock continued her career in cytogenetics studying the mechanics and inheritance of broken and ring (circular) chromosomes of maize. During her cytogenetic work, McClintock discovered transposons, a find which eventually led to her Nobel Prize in 1983.

Natural Populations of Drosophila

In the 1930s, Dobzhansky and his coworkers collected *Drosophila pseudoobscura* and *D. persimilis* from wild populations in California and neighbouring states. Using Painter's technique they studied the polytene chromosomes and discovered that the wild populations were polymorphic for chromosomal inversions. All the flies look alike whatever inversions they carry: this is an example of a cryptic polymorphism.

Evidence rapidly accumulated to show that natural selection was responsible. Using a method invented by L'Heretier and Teissier, Dobzhansky bred populations in *population cages*, which enabled feeding, breeding and sampling whilst preventing escape. This had the benefit of eliminating migration as a possible explanation of the results. Stocks containing inversions at a known initial frequency can be maintained in controlled conditions. It was found that the various chromosome types do not fluctuate at random, as they would if selectively neutral, but adjust to certain frequencies at which they become stabilised. By the time Dobzhansky published the third edition of his book in 1951 he was persuaded that the chromosome morphs were being maintained in the population by the selective advantage of the heterozygotes, as with most polymorphisms.

Human Abnormalities and Medical Applications

In the event of procedures which allowed easy enumeration of chromosomes, discoveries were quickly made related to aberrant chromosomes or chromosome number. In some congenital disorders, such as Down syndrome, cytogenetics revealed the nature of the chromosomal defect: a "simple" trisomy. Abnormalities arising from nondisjunction events can cause cells with aneuploidy (additions or deletions of entire chromosomes) in one of the parents or in the fetus.

In 1959, Lejeune discovered patients with Down syndrome had an extra copy of chromosome 21. Down syndrome is also referred to as trisomy 21.

Other numerical abnormalities discovered include sex chromosome abnormalities. A female with only one X chromosome has Turner syndrome, an additional X chromosome in a male, resulting in 47 total chromosomes, has Klinefelter Syndrome. Many other sex chromosome combinations are compatible with live birth including XXX, XYY, and XXXX. The ability for mammals to tolerate aneuploidies in the sex chromosomes arises from the ability to inactivate them, which is required in normal females to compensate for having two copies of the chromosome. Not all genes on the X chromosome are inactivated, which is why there is a phenotypic effect seen in individuals with extra X chromosomes.

Trisomy 13 was associated with Patau Syndrome and trisomy 18 with Edwards Syndrome.

In 1960, Peter Nowell and David Hungerford discovered a small chromosome in the white blood cells of patients with Chronic myelogenous leukemia (CML). This abnormal chromosome was dubbed the Philadelphia chromosome - as both scientists were doing their research in Philadelphia, Pennsylvania. Thirteen years later, with the development of more advanced techniques, the abnormal chromosome was shown by Janet Rowley to be the result of a translocation of chromosomes 9 and 22. Identification of the Philadelphia chromosome by cytogenetics is diagnostic for CML.

Advent of Banding Techniques

In the late 1960s, Caspersson developed banding techniques which differentially stain chromosomes. This allows chromosomes of otherwise equal size to be differentiated as well as to elucidate the breakpoints and constituent chromosomes involved in chromosome translocations. Deletions within one chromosome could also now be more specifically named and understood.

Diagrams identifying the chromosomes based on the banding patterns are known as *idiograms*. These maps became the basis for both prenatal and oncological fields to quickly move cytogenetics into the clinical lab where karyotyping allowed scientists to look for chromosomal alterations. Techniques were expanded to allow for culture of free amniocytes recovered from amniotic fluid, and elongation techniques for all culture types that allow for higher-resolution banding.

Beginnings of Molecular Cytogenetics

In the 1980s, advances were made in molecular cytogenetics. While radioisotope-labelled probes had been hybridized with DNA since 1969, movement was now made in using fluorescent labelled probes. Hybridizing them to chromosomal preparations using existing techniques came to be known as fluorescence *in situ* hybridization (FISH). This change significantly increased the usage of probing techniques as fluorescent labelled probes are safer. Further advances in micromanipulation and examination of chromosomes led to the technique of chromosome microdissection whereby aberrations in chromosomal structure could be isolated, cloned and studied in ever greater detail.

Techniques

Karyotyping: Routine chromosome analysis (Karyotyping) refers to analysis of metaphase chromosomes which have been banded using trypsin followed by Giemsa, Leishmanns, or a mixture of the two. This creates unique banding patterns on the chromosomes. The molecular mechanism and reason for these patterns is unknown, although it likely related to replication timing and chromatin packing.

Several chromosome-banding techniques are used in cytogenetics laboratories. Quinacrine banding (Q-banding) was the first staining method used to produce specific banding patterns. This method requires a fluorescence microscope and is no longer as widely used as Giemsa banding (G-banding). Reverse banding, or R-banding, requires heat treatment and reverses the usual black-and-white pattern that is seen in G-bands and Q-bands.

This method is particularly helpful for staining the distal ends of chromosomes. Other staining techniques include C-banding and nucleolar organizing region stains (NOR stains). These latter methods specifically stain certain portions of the chromosome. C-banding stains the constitutive heterochromatin, which usually lies near the centromere, and NOR staining highlights the satellites and stalks of acrocentric chromosomes. High-resolution banding involves the staining of chromosomes during prophase or early metaphase (prometaphase), before they reach maximal condensation. Because prophase and prometaphase chromosomes are more extended than metaphase chromosomes, the number of bands observable for all chromosomes increases from about 300 to 450 to as many as 800. This allows the detection of less obvious abnormalities usually not seen with conventional banding.

Slide Preparation

Cells from bone marrow, blood, amniotic fluid, cord blood, tumor, and tissues (including skin, umbilical cord, chorionic villi, liver, and many other organs) can be cultured using standard cell culture techniques in order to increase their number. A mitotic inhibitor (colchicine, colcemid) is then added to the culture. This stops cell division at mitosis which allows an increased yield of mitotic cells for analysis. The cells are then centrifuged and media and mitotic inhibitor are removed, and replaced with a hypotonic solution. This causes the white blood cells or fibroblasts to swell so that the chromosomes will spread when added to a slide as well as lyses the red blood cells. After the cells have been allowed to sit in hypotonic solution, Carnoy's fixative (3:1 methanol to glacial acetic acid) is added. This kills the cells and hardens the nuclei of the remaining white blood cells. The cells are generally fixed repeatedly to remove any debris or remaining red blood cells. The cell suspension is then dropped onto specimen slides. After aging the slides in an oven or waiting a few days they are ready for banding and analysis.

Analysis

Analysis of banded chromosomes is done at a microscope by a clinical laboratory specialist in cytogenetics (CLSp(CG)). Generally 20 cells are analyzed which is enough to rule out mosaicism to an acceptable level. The results are summarized and given to a board-certified cytogeneticist for review, and to write an interpretation taking into account the patients previous history and other clinical findings. The results are then given out reported in an *International System for Human Cytogenetic Nomenclature 2009* (ISCN2009).

Fluorescent in Situ Hybridization

Fluorescent in situ hybridization refers to using fluorescently labelled probe to hybridize to cytogenetic cell preparations. In addition to standard preparations FISH can also be performed on:

- bone marrow smears
- blood smears
- paraffin embedded tissue preparations
- enzymatically dissociated tissue samples
- uncultured bone marrow
- uncultured amniocytes
- cytospin preparations

Slide Preparation

The slide is aged using a salt solution usually consisting of 2X SSC (salt, sodium citrate). The slides are then dehydrated in ethanol, and the probe mixture is added. The sample DNA and the probe DNA are then co-denatured using a heated plate and allowed to re-anneal for at least 4 hours. The slides are then washed to remove excess unbound probe, and counterstained with 4',6-Diamidino-2-phenylindole (DAPI) or propidium iodide.

Analysis

Analysis of FISH specimens is done by fluorescence microscopy by a clinical laboratory specialist in cytogenetics. For oncology generally a large number of interphase cells are scored in order to rule out low-level residual disease, generally between 200 and 1,000 cells are counted and scored. For congenital problems usually 20 metaphase cells are scored.

Future of Cytogenetics

Advances now focus on molecular cytogenetics including automated systems for counting the results of standard FISH preparations and techniques for virtual karyotyping, such as comparative genomic hybridization arrays, CGH and Single nucleotide polymorphism arrays.

Carbohydrates

Carbohydrates, which include cellulose, starches, sugars, and many other compounds, are the most abundant single class of organic substances found in nature. They are formed in green plants and certain bacteria by a process known as photosynthesis, in which energy derived from sunlight is used for the assimilation of carbon dioxide from the air. If carbon dioxide, water, minerals, and an appropriate inorganic source of nitrogen are available, these organisms, with the aid of solar energy, can synthesize all the different carbohydrates, proteins, and lipids they need for their existence. Other organisms cannot do this. It follows that life on Earth ultimately depends on this process of carbon dioxide assimilation in which carbohydrates are the first intermediates.

Chemists in the 19th century found that carbohydrates contain the elements of carbon, hydrogen, and oxygen. Hence they referred to them as carbon hydrates, the contracted form of which is still used today, even though it is now known that deviations from the required hydrogen-to-oxygen ratio of water do occur.

Occurrence and Function

Carbohydrates function as the main structural elements in plants, in two forms: cellulose, a polymer of glucose, and hemicelluloses, which are polymers of 5-carbon sugars and other compounds. Carbohydrates serve as storage products of energy. The principal forms are starch in plants and glycogen in animal tissues. These are polymers of glucose; they are deposited in cells in the form of granules when a surplus of glucose is available. In times of metabolic need, when the body is exerting itself, the polymers are broken down by enzymatic action and become fuel.

Plants store starch in roots, tubers, and leafy parts, mainly during photosynthetic activity; some plants, such as sugar beets and sugarcane, also store sucrose. A large part of the human diet consists of carbohydrates in the form of starch and sucrose. Both must first be broken down to their component sugars by digestive enzymes before absorption into the bloodstream can take place.

Carbohydrate Chemistry

Single units of carbohydrates known as monosaccharides, or simple sugars, are compounds with either an aldehyde group (aldose sugars) or a ketone group (ketose sugars). Monosaccharides are classified as trioses, tetroses, pentoses, and so on, indicating the number of carbon atoms in a molecule. The monosaccharide series begins with glyceraldehyde, an aldotriose with one asymmetric carbon atom.

Disaccharides and polysaccharides are formed from two or more monosaccharides joined by chemical bonds. Glucose linked to fructose, for instance, forms the disaccharide sucrose (cane sugar), and glucose linked to galactose forms the disaccharide lactose (milk sugar). Starch, glycogen, and cellulose are all chains of glucose units, differing only in their modes of bonding and degree of chain branching. Some biologically important sugar derivatives are sugar alcohols, sugar acids, deoxy sugars, amino sugars, sugar phosphates, and muramic and neuraminic acids.

Metabolism

Photosynthesis, through the operation of the food chain, is the ultimate source of energy for nearly all organisms. Furthermore, the fossil fuels (coal, petroleum, and natural gas) and the oxygen of the Earth's atmosphere are all derived from earlier activity of photosynthetic organisms. In green plants, with the aid of chlorophyll and other pigment, light energy is transformed into chemical energy.

The reaction involves water and carbon dioxide converted into glucose and oxygen. This reaction may be divided into a light-dependent reaction, in which chemical energy is created, and a light-independent reaction (dark reaction), in which carbon dioxide is used to make carbohydrate at the expense of the chemical energy created in the light reaction

. The utilization of carbohydrate by oxidation occurs in a reaction that is the reverse of that of photosynthesis. In plants the glucose is stored as starch and sucrose. In humans excess glucose is stored as glycogen and also is converted to fat and stored in adipose tissues. Carbohydrate metabolic processes such as these are subject to regulatory controls in which various hormones play a predominant role.

Equisetum

Equisetum is the only living genus in Equisetaceae, a family of vascular plants that reproduce by spores rather than seeds.

Equisetum is a "living fossil" as it is the only living genus of the entire class Equisetopsida, which for over one hundred million years was much more diverse and dominated the understory of late Paleozoic forests. Some Equisetopsida were large trees reaching to 30 metres tall. The genus *Calamites* of the family Calamitaceae, for example, is abundant in coal deposits from the Carboniferous period.

A superficially similar but entirely unrelated flowering plant genus, mare's tail (*Hippuris*), is occasionally misidentified as "horsetail".

It has been suggested that the pattern of spacing of nodes in horsetails, wherein those toward the apex of the shoot are increasingly close together, inspired John Napier to discover logarithms.

Etymology

The small white protuberances are accumulated silicates on cells.

The name "horsetail", often used for the entire group, arose because the branched species somewhat resemble a horse's tail. Similarly, the scientific name *Equisetum* derives from the Latin *equus* ("horse") + *seta* ("bristle").

Other names include candock for branching individuals, and snake grass or scouring-rush for unbranched or sparsely branched individuals. The latter name refers to the rush-like appearance appearance of the plants, and to the fact that the stems are coated with abrasive silicates, making them useful for scouring (cleaning) metal items such as cooking pots or drinking mugs, particularly those made of tin. In German, the corresponding name is *Zinnkraut* ("tin-herb"). Rough horsetail *E. hyemale* is still boiled and then dried in Japan, to be used for the final polishing process on woodcraft to produce a smoother finish than any sandpaper.

Distribution, Ecology and Uses

The genus *Equisetum* is near-cosmopolitan, being absent only from Antarctica, though they are not known to be native to Australia, New Zealand nor the islands of the Pacific. They are perennial plants, either herbaceous and dying back in winter as most temperate species, or evergreen as most tropical species and the temperate species rough horsetail (*E. hyemale*), branched horsetail (*E. ramosissimum*), dwarf horsetail (*E. scirpoides*) and variegated horsetail (*E. variegatum*). They typically grow 0.2-1.5 m tall, though the "giant horsetails" are recorded to grow as high as 2.5 m (northern giant horsetail, *E. telmateia*), 5 m (southern giant horsetail, *E. giganteum*) or 8 m (Mexican giant horsetail, *E. myriochaetum*), and allegedly even more.

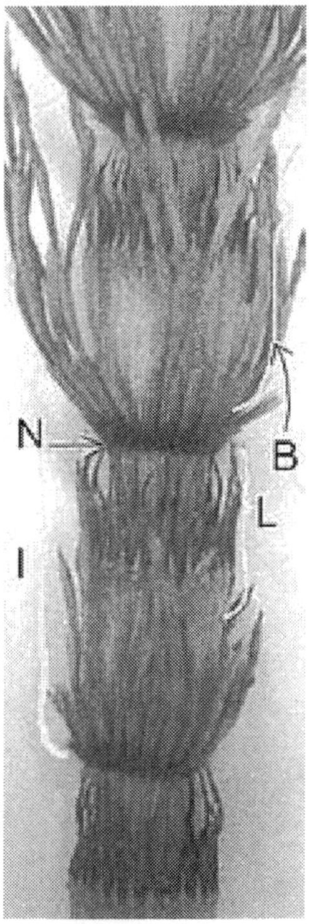

Figure: *Vegetative stem: B = branch in whorl I = internode L = leaves N = node*

Many species in this genus prefer wet sandy soils, though some are semi-aquatic and others are adapted to wet clay soils. The stalks arise from rhizomes that are deep underground and almost impossible to dig out. The field horsetail (*E. arvense*) can be a nuisance weed, readily regrowing from the rhizome after being pulled out. It is also unaffected by many herbicides designed to kill seed plants. However, as *E. arvense* prefers an acid soil, lime may be used to assist in eradication efforts to bring the soil pH to 7 or 8. Members of the genus have been declared noxious weeds in Australia and in the US state of Oregon.

All the *Equisetum* are classed as "unwanted organisms" in New Zealand and are listed on the National Pest Plant Accord.

If eaten over a long enough period of time, some species of horsetail can be poisonous to grazing animals, including horses. The toxicity appears to be due to thiaminase enzymes, which can cause thiamine deficiency. People have regularly consumed horsetails. The young plants are eaten cooked or raw, but considerable care must be taken. For example, the fertile stems bearing strobili of some species are cooked and eaten like asparagus (a dish called *tsukushi*) in Japan. The people of ancient Rome would eat meadow horsetail in a similar manner, but they also used it to make tea as well as a thickening powder. Indians of the North American Pacific Northwest eat the young shoots of this plant raw. The plants are used as a dye and give a soft green colour. An extract is often used to provide silica for supplementation. Horsetail was often used by Indians to polish wooden tools. *Equisetum* species are often used to analyze gold concentrations in an area due to their ability to take up the metal when it is in a solution.

Anatomy

In these plants the leaves are greatly reduced and usually non-photosynthetic. They contain a single, non-branching vascular trace, which is the defining feature of microphylls. However, it has recently been recognised that horsetail microphylls are probably not ancestral as in Lycopodiophyta (clubmosses and relatives), but rather derived adaptations, evolved by reduction of megaphylls. They are, therefore, sometimes actually referred to as megaphylls to reflect this homology.

The leaves of horsetails are arranged in whorls fused into nodal sheaths. The stems are green and photosynthetic, and are distinctive in being hollow, jointed and ridged (with sometimes 3 but usually 6-40 ridges). There may or may not be whorls of branches at the nodes.

Figure: *Strobilus of Northern Giant Horsetail (*Equisetum telmateia braunii*), terminal on an unbranched stem.*

Spores

The spores are borne under sporangiophores in strobili, cone-like structures at the tips of some of the stems. In many species the cone-bearing shoots are unbranched, and in some (e.g. field horsetail, *E. arvense*) they are non-photosynthetic, produced early in spring separately from photosynthetic, sterile shoots. In some other species (e.g. marsh horsetail, *E. palustre*) they are very similar to sterile shoots, photosynthetic and with whorls of branches.

Horsetails are mostly homosporous, though in the field horsetail smaller spores give rise to male prothalli. The spores have four elaters that act as moisture-sensitive springs, assisting spore dispersal after the sporangia have split open longitudinally.

Systematics

Species: The living members of the genus *Equisetum* are divided into two distinct lineages, which are usually treated as subgenera.

Hybrids are common, but hybridization has only been recorded between members of the same subgenus.

In addition, there are numerous ill-determined populations. One of them, the Kamchatka Horsetail (*Equisetum camtschatcense*), is an ornamental forming imposing stands of these archaic plants.

Subgenus *Equisetum*

- *Equisetum arvense* L. – Field Horsetail, Common Horsetail or Giant Horsetail
- *Equisetum bogotense* Kunth – Andean Horsetail or Giant Horsetail
- *Equisetum diffusum* L. – Himalayan Horsetail
- *Equisetum fluviatile* L. – Water Horsetail
- *Equisetum palustre* L. – Marsh Horsetail
- *Equisetum pratense* Ehrh. – Meadow Horsetail, Shade Horsetail, Shady Horsetail
- *Equisetum sylvaticum* L. – Wood Horsetail
- *Equisetum telmateia* Ehrh. – Great Horsetail, Northern Giant Horsetail

Subgenus *Hippochaete*

- *Equisetum giganteum* L. – Southern Giant Horsetail or Giant Horsetail
- *Equisetum myriochaetum* Schltdl. & Cham. – Mexican Giant Horsetail or Giant Horsetail
- *Equisetum hyemale* L. – Rough Horsetail, Scouringrush Horsetail
- *Equisetum laevigatum* A. Braun – Smooth Horsetail
- *Equisetum ramosissimum* Desf. – Branched Horsetail
- *Equisetum scirpoides* Michx. – Dwarf Horsetail
- *Equisetum variegatum* Schleich. ex Weber & Mohr – Variegated Horsetail

unplaced to subgenus

- †*Equisetum thermale* - Middle to Late Jurassic, Argentina

Named Hybrids

Hybrids between species in subgenus *Equisetum*

- *Equisetum* × *bowmanii* C.N.Page (*Equisetum sylvaticum* × *Equisetum telmateia*)

- *Equisetum* × *dycei* C.N.Page (*Equisetum fluviatile* × *Equisetum palustre*)
- *Equisetum* × *font-queri* Rothm. (*Equisetum palustre* × *Equisetum telmateia*)
- *Equisetum* × *litorale* Kühlew ex Rupr. (*Equisetum arvense* × *Equisetum fluviatile*)
- *Equisetum* × *mchaffieae* C.N.Page (*Equisetum fluviatile* × *Equisetum pratense*)
- *Equisetum* × *mildeanum* Rothm. (*Equisetum pratense* × *Equisetum sylvaticum*)
- *Equisetum* × *robertsii* Dines (*Equisetum arvense* × *Equisetum telmateia*)
- *Equisetum* × *rothmaleri* C.N.Page (*Equisetum arvense* × *Equisetum palustre*)
- *Equisetum* × *willmotii* C.N.Page (*Equisetum fluviatile* × *Equisetum telmateia*)

Hybrids between species in subgenus *Hippochaete*

- *Equisetum* × *ferrissii* Clute (*Equisetum hyemale* × *Equisetum laevigatum*)
- *Equisetum* × *moorei* Newman (*Equisetum hyemale* × *Equisetum ramosissimum*)
- *Equisetum* × *nelsonii* (A.A.Eaton) Schaffn. (*Equisetum laevigatum* × *Equisetum variegatum*)
- *Equisetum* × *schaffneri* Milde (*Equisetum giganteum* × *Equisetum myriochaetum*)
- *Equisetum* × *trachydon* A.Braun (*Equisetum hyemale* × *Equisetum variegatum*)

Equisetum *Cell Walls*

The crude cell extracts of all *Equisetum* species tested contain mixed-linkage glucan : Xyloglucan endotransglucosylase (MXE) activity. This is a novel enzyme and is not known to occur in any other plants. In addition, the cell walls of all *Equisetum* species tested contain mixed-linkage glucan (MLG), a polysaccharide which, until recently, was thought to be confined to the Poales. The evolutionary distance between *Equisetum* and the Poales suggests that each evolved MLG independently. The presence of MXE activity in *Equisetum* suggests that they have evolved MLG along with some mechanism of cell wall modification. The lack of MXE in the Poales suggests that there it must play some other, currently unknown, role. Due to the correlation between MXE activity and cell age, MXE has been proposed to promote the cessation of cell expansion.

Medicinal Uses

The plant has a long history of medicinal uses, although modern sources include cautions with regard to its use. In 2009 the European Food Safety Authority issued a report assessing health claims (invigoration, weight control, and skin, hair, and bone health) for *Equisetum arvense L.*, concluding that none could be substantiated.

Angiopteris

Angiopteris is a genus of huge evergreen ferns from the Marattiaceae family, found throughout the paleotropics from Madagascar to the South Pacific islands. Species of smaller stature with elongate synangia and creeping rhizomes are sometimes segregated into the genus *Archangiopteris*, and a once-pinnate monotypic segregate genus has been called *Macroglossum*, but molecular data supports inclusion of these taxa within a broad concept of *Angiopteris*.

Angiopteris evecta has been introduced and naturalized in Hawaii, Jamaica, and parts of Central America, where it has become an invasive weed in lower elevation drainages. They feature a large, erect, woody rhizome with a wide base supported by thick roots. The fronds are deltoid, pinnate, 5–8 metres (16–26 ft) long, with spreading leaflets.

The basal chromosome number for this genus is $2n=80$. The type species is *Angiopteris evecta*.

Angiopteris Evecta

Angiopteris evecta, commonly known as the Giant Fern, is a rare plant occurring in eastern and northern Australia. Also found growing in nearby islands such as New Guinea and various places in Polynesia and Melanesia. Listed as endangered in New South Wales, where it has been recorded growing in sub tropical rainforest, in the valley of the Tweed River. It is an invasive species in Jamaica.

Angiopteris evecta is the type species of the genus *Angiopteris*. It was originally described as *Polypodium evectum* by Georg Forster in 1786, before being reclassified and given its current binomial name in 1796 by Georg Franz Hoffmann. The species name is the Latin adjective *evectus* "swollen" or "inflated". Common names include giant fern, king fern, oriental vessel fern, and mule's foot fern. The huge mature fronds measure up to 8 metres (25 ft) long. They originate from a large thick rootstock, up to 150 cm (60 in) high.

Angiopteris evecta can be grown in well-drained moist sites in the garden with some shade. It is unable to be propagated by spores but the lobes from the frond base can be removed and will form a new plant in around a year in a medium of sand and peat.

Salviniales

The order Salviniales (formerly known as the Hydropteridales and including the former Marsileales) is an order of ferns in the class Polypodiopsida. Salviniales are all aquatic and differ from all other ferns in being heterosporous, meaning that they produce two different types of spores (megaspores and microspores) that develop into two different types of gametophytes (female and male gametophytes, respectively), and in that their gametophytes are endosporic, meaning that they never grow outside the spore wall and cannot become larger than the spores that produced them. In being heterosporus with endosporic gametophytes they are more similar to seed plants than to other ferns.

The ferns of this order vary radically in form from one another and do not look particularly fern-like. Species of the family Salviniaceae are natant (floating), while those of the family Marsileaceae are rooted. However, the natant species may temporarily grow on wet mud during times of low water, and the Marsileaceae may grow as emergent species, depending on species and location.

There is a well-known fossil member of the Marsileales, *Hydropteris* (*incertae sedis*).

Azolla

Azolla (mosquito fern, duckweed fern, fairy moss, water fern) is a genus of seven species of aquatic ferns in the family Salviniaceae. They are extremely reduced in form and specialized, looking nothing like conventional ferns but more resembling duckweed or some mosses.

Selected Species

Azolla caroliniana Willd.

[Poss. synon. of *A. filiculoides*]

Azolla circinata Oltz & Hall

Azolla filiculoides Lam.

Azolla japonica Franch. & Sav.

[Poss. synon. of *A. filiculoides*]

Azolla mexicana C.Presl

Azolla microphylla Kaulf.

Azolla nilotica Decne. ex Mett.

Azolla pinnata R.Br.

Azolla rubra R.Br.

[Poss. synon. of *A. filiculoides*]

List sources :

At least six extinct species are known from the fossil record:

- *Azolla intertrappea* Sahni & H.S. Rao, 1934 (Eocene, India)
- *Azolla berryi* Brown, 1934 (Eocene, Green River Formation, Wyoming)
- *Azolla prisca* Chandler & Reid, 1926 (Oligocene, London Clay, Isle of Wight)
- *Azolla tertiaria* Berry, 1927 (Pliocene, Esmeralda Formation, Nevada)
- *Azolla primaeva* (Penhallow) Arnold, 1955 (Eocene, Allenby Formation, British Columbia)
- *Azolla boliviensis* Vajda & McLoughlin, 2005 (Maastrichtian - Paleocene, Eslaboacuten Formation& Flora Formation Bolvia)

Ecology

Azolla filiculoides (Red Azolla) is the only member of this genus and of the family Azollaceae in Tasmania. It is a very common native aquatic plant in Tasmania. It is particularly common on farm dams and other still water bodies. The plants are small (usually only a few cm across) and float, but can be very abundant and form large mats. The plants are typically red, and have very small water repellent leaves. *Azolla* floats on the surface of water by means of numerous, small, closely overlapping scale-like leaves, with their roots hanging in the water. They form a symbiotic relationship with the cyanobacterium *Anabaena azollae,* which fixes atmospheric nitrogen, giving the plant access to the essential nutrient. This has led to the plant being dubbed a "super-plant", as it can readily colonise areas of freshwater, and grow at great speed - doubling its biomass every two to three days. The only known limiting factor on its growth is phosphorus, another essential mineral. An abundance of phosphorus, due for example to eutrophication or chemical runoff, often leads to *Azolla* blooms.

The nitrogen-fixing capability of *Azolla* has led to *Azolla* being widely used as a biofertilizer, especially in parts of southeast Asia. Indeed, the plant has been used to bolster agricultural productivity in China for over a thousand years. When rice paddies are flooded in the spring, they can be inoculated with *Azolla*, which then quickly multiplies to cover the water, suppressing weeds. The rotting plant material releases nitrogen to the rice plants, providing up to nine tonnes of protein per hectare per year.

Azolla are also serious weeds in many parts of the world, entirely covering some bodies of water. The myth that no mosquito can penetrate the coating of fern to lay its eggs in the water gives the plant its common name "mosquito fern". Most of the species can produce large amounts of deoxyanthocyanins in response to various stresses, including bright sunlight and extremes of temperature, causing the water surface to appear to be covered with an intensely red carpet. Herbivore feeding induces accumulation of deoxyanthocyanins and leads to a reduction in the proportion of polyunsaturated fatty acids in the fronds, thus lowering their palatability and nutritive value.

Azolla cannot survive winters with prolonged freezing, so is often grown as an ornamental plant at high latitudes where it cannot establish itself firmly enough to become a weed. It is not tolerant to salinity; normal plants can't survive in greater than 1-1.6‰, and even conditioned organisms die in over 5.5‰ salinity.

Reproduction

Azolla reproduces sexually, and asexually by splitting.

Like all ferns, sexual reproduction leads to spore formation, but *Azolla* sets itself apart from other members of its group by producing two kinds. During the summer months, numerous spherical structures called sporocarps form on the undersides of the branches. The male sporocarp is greenish or reddish and looks like the egg mass of an insect or spider. It is two millimeters in diameter, and inside are numerous male sporangia. Male spores (microspores) are extremely small and are produced inside each microsporangium. Curiously, microspores tend to adhere in clumps called massulae.

Female sporocarps are much smaller, containing one sporangium and one functional spore. Since an individual female spore is considerably larger than a male spore, it is termed a megaspore.

Azolla has microscopic male and female gametophytes that develop inside the male and female spores. The female gametophyte protrudes from the megaspore and bears a small number of archegonia, each containing a single egg. The microspore forms a male gametophyte with a single antheridium which produces eight swimming sperm. The barbed glochidia on the male spore clusters are assumed to cause them to cling to the female megaspores, thus facilitating fertilization.

Human Use

Food: In addition to its traditional cultivation as a bio-fertilizer for wetland paddy (due to its ability to fix nitrogen), azolla is finding increasing use for sustainable production of livestock feed. Azolla is rich in proteins, essential amino acids, vitamins and minerals. Studies describe feeding azolla to dairy cattle, pigs, ducks, and chickens, with reported increases in milk production, weight of broiler chickens and egg production of layers, as compared to conventional feed. One FAO study describes how azolla integrates into a tropical biomass agricultural system, reducing the need for inputs. Azolla has also been suggested as a food stuff for human consumption. However, no long term studies of the healthiness of eating Azolla have been made on humans.

Companion Plant

Azolla has been used, for at least one thousand years, in rice paddies as a companion plant, because of its ability to both fix nitrogen, and block out light to prevent any competition from other plants, aside from the rice, which is planted when tall enough to poke out of the water through the azolla layer. Mats of mature azolla can also be used as a weed-suppressing mulch.

Larvicide

As an additional benefit to its role as a paddy biofertilizer, *Azolla* spp. have been used to control mosquito larvae in rice fields. The plant grows in a thick mat on the surface of the water, making it more difficult for the larvae to reach the surface to breathe, effectively choking the larvae.

Importance in Paleoclimatology

A study of Arctic Paleoclimatology reported that *Azolla* may have had a significant role in reversing an increase in greenhouse effect that occurred 55 million years ago that caused the region around the north pole to turn into a hot, tropical environment. This research conducted by the Institute of Environmental Biology at Utrecht

University claims that massive patches of *Azolla* growing on the (then) freshwater surface of the Arctic Ocean consumed enough carbon dioxide from the atmosphere for the global greenhouse effect to decline, eventually causing the formation of Ice sheets in Antarctica and the current "Icehouse period" which we are still in. This theory has been termed the Azolla event.

Salvinia

Salvinia, a genus in the family Salviniaceae, is a floating fern named in honour of Antonio Maria Salvini, a 17th-century Italian scientist. The genus was published by Séguier, in Pl. Veron. 3: 52. 1754. 12 species are currently recognized, at least three of which (*S. molesta*, *S. herzogii*, and *S. minima*) are believed to be hybrids.

Salvinia is related to the other water ferns, including the Mosquito fern *Azolla*. Recent sources include both *Azolla* and *Salvinia* in Salviniaceae, although each genus was formerly given its own family.

Salvinia, like the other ferns in order Salviniales are heterosporous, producing spores of differing sizes. However, leaf development in *Salvinia* is unique. The upper side of the floating leaf, which appears to face the stem axis, is morphologically abaxial (J. G. Croxdale 1978, 1979, 1981).

The plants pose a particular hindrance on certain lakes, having choked off much of the water in Lake Bistineau near Doyline in Webster Parish, Louisiana. Other problems have persisted in a second Webster Parish site, Caney Lakes Recreation Area.

Distribution

Mostly tropical, North America, Mexico, West Indies, Central America, South America, Eurasia and Africa, including Madagascar.

Description

Small, floating aquatics with creeping stems, branched, bearing hairs but no true roots. Leaves in whorls of 3, with 2 leaves green, sessile or short-petioled, flat, entire, and floating, 1 leaf finely dissected, petiolate, rootlike, and pendent. Submerged leaves bearing sori that are surrounded by basifixed membranous indusia (sporocarps); sporocarps of 2 types, bearing either megasporangia that are few in number (ca. 10), each with single megaspore, or many microsporangia, each with 64 microspores. Spores of 2 kinds and sizes, both globose, trilete. Megagametophytes and microgametophytes protruding through sporangium wall; megagametophytes floating on water surface with archegonia directed downward; microgametophytes remaining fixed to sporangium wall. The small, hairlike growths, known as microgametical follicles, are not known to have any productive function, and are currently a biological mystery.

Economic Uses

Giant salvinia (*Salvinia molesta*) is a commonly introduced invasive weed in warm climates. It grows rapidly and forms dense mats over still waters. It is native to South America. A tiny weevil, *Cyrtobagous salviniae*, has been used successfully to control giant salvinia.

Prokaryote

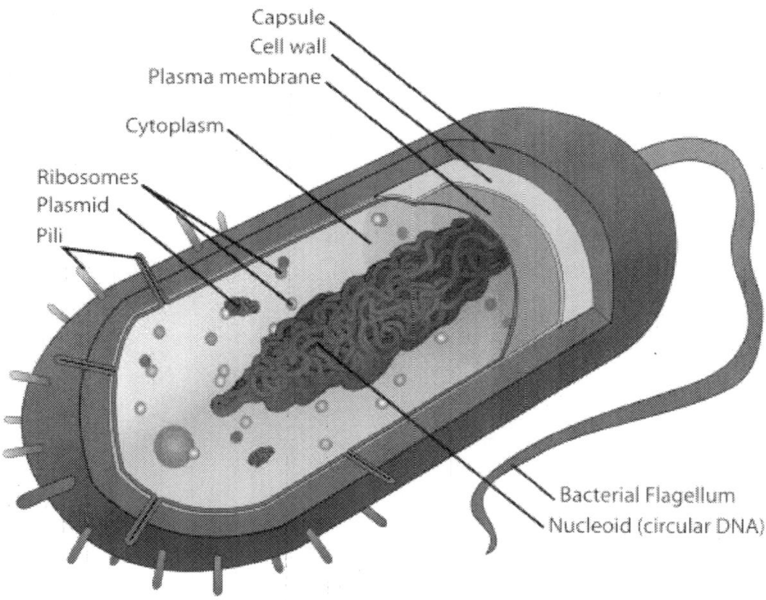

Figure: *Cell structure of a bacterium, a member of one of the two domains of prokaryotic life*

The prokaryotes are a group of organisms whose cells lack a membrane-bound nucleus (karyon). The organisms whose cells do have a nucleus are called eukaryotes. Most prokaryotes are unicellular organisms, although a few such as myxobacteria have multicellular stages in their life cycles or create large colonies like cyanobacteria. The word *prokaryote* comes from the Greek *"pro'* meaning "before" and *karyon* meaning "nut or kernel". Prokaryotes do not have a nucleus, mitochondria, or any other membrane-bound organelles. In other words, all their intracellular water-soluble components (proteins, DNA and metabolites) are located together in the same volume enclosed by the cell membrane, rather than in separate cellular compartments.

The division to prokaryotes and eukaryotes reflects two distinct levels of cellular organization rather than biological classification of species. Prokaryotes include two major classification domains: the bacteria and the archaea. Archaea were recognized as a domain of life in 1990. These organisms were originally thought to live only in inhospitable conditions such as extremes of temperature, pH, and radiation but have since been found in all types of habitats.

Relationship to Eukaryotes

The division between prokaryotes and eukaryotes is usually considered the most important distinction or difference among organisms. The distinction is that eukaryotic cells have a "true" nucleus containing their DNA, whereas prokaryotic cells do not have a nucleus. One criticism of this classification points out that the word "prokaryote" is based on what these organisms are not (they are not eukaryotic), rather than what they are (either archaea or bacteria). Another difference is that ribosomes in prokaryotes are smaller than in eukaryotes. However, two organelles found in many eukaryotic cells, mitochondria and chloroplasts, contain ribosomes similar in size and makeup to those found in prokaryotes. This is one of many pieces of evidence that mitochondria and chloroplasts are themselves descended from free-living bacteria. This theory holds that early eukaryotic cells took in primitive prokaryotic cells by phagocytosis and adapted themselves to incorporate their structures, leading to the mitochondria we see today.

The genome in a prokaryote is held within a DNA/protein complex in the cytosol called the nucleoid, which lacks a nuclear envelope. The complex contains a single, cyclic, double-stranded molecule of stable chromosomal DNA, in contrast to the multiple linear, compact, highly organized chromosomes found in eukaryotic cells. In addition, many important genes of prokaryotes are stored in separate circular DNA structures called plasmids.

Prokaryotes lack mitochondria and chloroplasts. Instead, processes such as oxidative phosphorylation and photosynthesis take place across the prokaryotic cell membrane. However, prokaryotes do possess some internal structures, such as prokaryotic cytoskeletons, and the bacterial order Planctomycetes have a membrane around their nucleoid and contain other membrane-bound cellular structures. Both eukaryotes and prokaryotes contain large RNA/protein structures called ribosomes, which produce protein. Prokaryotic cells are usually much smaller than eukaryotic cells. Therefore, prokaryotes have a larger surface-area-to-volume ratio, giving them a higher metabolic rate, a higher growth rate, and as a consequence, a shorter generation time than eukaryotes.

In 1977, Carl Woese proposed dividing prokaryotes into the Bacteria and Archaea (originally Eubacteria and Archaebacteria) because of the major differences in the structure and genetics between the two groups of organisms. This arrangement of Eukaryota (also called "Eukarya"), Bacteria, and Archaea is called the three-domain system, replacing the traditional two-empire system.

Sociality

While prokaryotes are considered to be strictly unicellular, most are capable of forming stable aggregate communities. When such communities are encased in a stabilizing polymer matrix ("slime"), they may be called "biofilms". Cells in biofilms often show distinct patterns of gene expression (phenotypic differentiation) in time and space. Also, as with multicellular eukaryotes, these changes in expression appear to often result from cell-to-cell signalling, a phenomenon known as quorum sensing.

Biofilms may be highly heterogeneous and structurally complex and may attach to solid surfaces, or exist at liquid-air interfaces, or potentially even liquid-liquid interfaces. Bacterial biofilms are often made up of microcolonies (approximately dome-shaped masses of bacteria and matrix) separated by "voids" through which the medium (e.g., water) may flow relatively uninhibited. The microcolonies may join together above the substratum to form a continuous layer, closing the network of channels separating microcolonies. This structural complexity—combined with observations that oxygen limitation (a ubiquitous challenge for anything growing in size beyond the scale of diffusion) is at least partially eased by movement of medium throughout the biofilm—has led some to speculate that this may constitute a circulatory system and many researchers have started calling prokaryotic communities multicellular (for example). Differential cell expression, collective behaviour, signalling, programmed cell death, and (in some cases) discrete biological dispersal events all seem to point in this direction. However, these colonies are seldom if ever founded by a single founder (in the way that animals and plants are founded by single cells), which presents a number of theoretical issues. Most explanations of co-operation and the evolution of multicellularity have focused on high relatedness between members of a group (or colony, or whole organism). If a copy of a gene is present in all members of a group, behaviours that promote cooperation between members may permit those members to have (on average) greater fitness than a similar group of selfish individuals.

Should these instances of prokaryotic sociality prove to be the rule rather than the exception, it would have serious implications for the way we view prokaryotes in general and the way we deal with them in medicine. Bacterial biofilms may be 100 times more resistant to antibiotics than free-living unicells and may be nearly impossible to remove from surfaces once they have colonized them. Other aspects of bacterial cooperation—such as bacterial conjugation and quorum-sensing-mediated pathogenicity—present additional challenges to researchers and medical professionals seeking to treat the associated diseases.

Reproduction

Bacteria and archaea reproduce through asexual reproduction, usually by binary fission. Genetic exchange and recombination still occur, but this a form of horizontal gene transfer and is not a replicative process, simply involving the transference of DNA between two cells, as in bacterial conjugation.

DNA Transfer

DNA transfer between prokaryotic cells occurs both in bacteria and archaea, although it has been mainly studied in bacteria. In bacteria, gene transfer occurs by three processes. These are

(1) bacterial virus (bacteriophage)-mediated transduction,

(2) plasmid-mediated conjugation, and

(3) natural transformation.

Transduction of bacterial genes by bacteriophage appears to reflect an occasional error during intracellular assembly of virus particles, rather than an adaptation of the host bacteria. The transfer of bacterial DNA is under the control of the bacteriophage's genes rather than bacterial genes. Conjugation in the well-studied *E. coli* system is controlled by plasmid genes, and is an adaptation for distributing copies of a plasmid from one bacterial host to another. Infrequently during this process, a plasmid may integrate into the host bacterial chromosome, and subsequently transfer part of the host bacterial DNA to another bacterium. Plasmid mediated transfer of host bacterial

DNA (conjugation) also appears to be an accidental process rather than a bacterial adaptation.

Natural bacterial transformation involves the transfer of DNA from one bacterium to another through the intervening medium. Unlike transduction and conjugation, transformation is clearly a bacterial adaptation for DNA transfer, because it depends on numerous bacterial gene products that specifically interact to perform this complex process. For a bacterium to bind, take up and recombine donor DNA into its own chromosome, it must first enter a special physiological state called competence. About 40 genes are required in *Bacillus subtilis* for the development of competence. The length of DNA transferred during *B. subtilis* transformation can be as much as a third to the whole chromosome. Transformation is a common mode of DNA transfer, and 67 prokaryotic species are thus far known to be naturally competent for transformation. The development of competence in nature is usually associated with stressful environmental conditions, and appears to be an adaptation for promoting repair of DNA damage in recipient cells.

Among archaea, *Halobacterium volcanii* forms cytoplasmic bridges between cells that appear to be used for transfer of DNA from one cell to another. Another archaeon, *Sulfolobus solfataricus*, transfers DNA between cells by direct contact. Frols et al. found that exposure of *S. solfataricus* to DNA damaging agents induces cellular aggregation, and suggested that cellular aggregation may enhance DNA transfer among cells to provide increased repair of damaged DNA via homologous recombination.

Structure

Recent research indicates that all prokaryotes actually do have cytoskeletons, albeit more primitive than those of eukaryotes. Besides homologues of actin and tubulin (MreB and FtsZ), the helically arranged building-block of the flagellum, flagellin, is one of the most significant cytoskeletal proteins of bacteria, as it provides structural backgrounds of chemotaxis, the basic cell physiological response of bacteria. At least some prokaryotes also contain intracellular structures that can be seen as primitive organelles. Membranous organelles (a.k.a. intracellular membranes) are known in some groups of prokaryotes, such as vacuoles or membrane systems devoted to special metabolic properties, e.g., photosynthesis or chemolithotrophy. In addition, some species also contain protein-enclosed microcompartments, which have distinct physiological roles (e.g., carboxysomes or gas vacuoles).

Most prokaryotes are between 1 μm and 10 μm, but they can vary in size from 0.2 μm to 750 μm (*Thiomargarita namibiensis*).

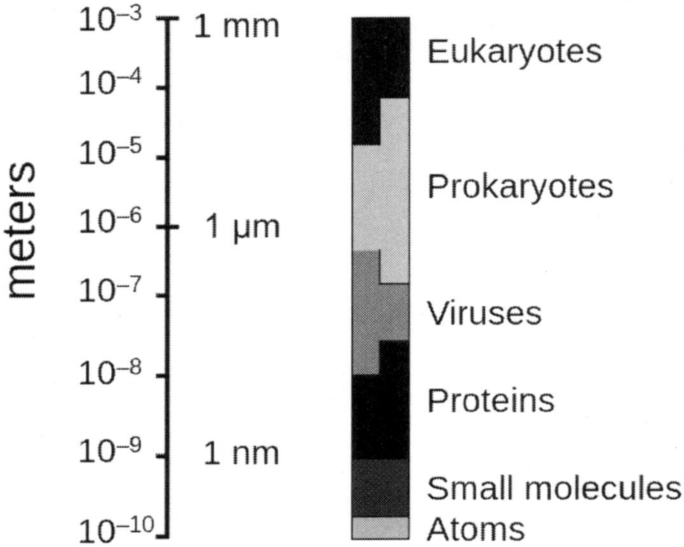

Figure: *The sizes of prokaryotes relative to other organisms and biomolecules*

Morphology of Prokaryotic Cells

Prokaryotic cells have various shapes; the four basic shapes are:

- Cocci – spherical
- Bacilli – rod-shaped
- Spirochaete – spiral-shaped
- Vibrio – comma-shaped

Environment

Prokaryotes live in nearly all environments on Earth. Some archaea and bacteria thrive in harsh conditions, such as high temperatures (thermophiles) or high salinity (halophiles). Organisms such as these are referred to as extremophiles. Many archaea grow as plankton in the oceans. Symbiotic prokaryotes live in or on the bodies of other organisms, including humans.

Evolution of Prokaryotes

The current model of the evolution of the first living organisms is that these were some form of prokaryotes, which may have evolved out of protobionts. In general, the eukaryotes are thought to have evolved later in the history of life. However, some authors have questioned this conclusion, arguing that the current set of prokaryotic species may have evolved from more complex eukaryotic ancestors through a process of simplification. Others have argued that the three domains of life arose simultaneously, from a set of varied cells that formed a single gene pool. This controversy was summarized in 2005:

There is no consensus among biologists concerning the position of the eukaryotes in the overall scheme of cell evolution. Current opinions on the origin and position of eukaryotes span a broad spectrum including the views that eukaryotes arose first in evolution and that prokaryotes descend from them, that eukaryotes arose contemporaneously with eubacteria and archeabacteria and hence represent a primary line of descent of equal age and rank as the prokaryotes, that eukaryotes arose through a symbiotic event entailing an endosymbiotic origin of the nucleus, that eukaryotes arose without endosymbiosis, and that eukaryotes arose through a symbiotic event entailing a simultaneous endosymbiotic origin of the flagellum and the nucleus, in addition to many other models, which have been reviewed and summarized elsewhere.

The oldest known fossilized prokaryotes were laid down approximately 3.5 billion years ago, only about 1 billion years after the formation of the Earth's crust. Eukaryotes only appear in the fossil record later, and may have formed from endosymbiosis of multiple prokaryote ancestors. The oldest known fossil eukaryotes are about 1.7 billion years old. However, some genetic evidence suggests eukaryotes appeared as early as 3 billion years ago.

While Earth is the only place in the universe where life is known to exist, some have suggested that there is evidence on Mars of fossil or living prokaryotes; but this possibility remains the subject of considerable debate and skepticism.

Prokaryotes have diversified greatly throughout their long existence. The metabolism of prokaryotes is far more varied than that of eukaryotes, leading to many highly distinct prokaryotic types. For example, in addition to using photosynthesis or organic compounds for energy, as eukaryotes do, prokaryotes may obtain energy from inorganic compounds such as hydrogen sulfide. This enables prokaryotes to thrive in harsh environments as cold as the snow surface of Antarctica, studied in cryobiology or as hot as undersea hydrothermal vents and land-based hot springs.

5

Bacterial Cell Structure

Bacteria, despite their simplicity, contain a well-developed cell structure which is responsible for many of their unique biological properties. Many structural features are unique to bacteria and are not found among archaea or eukaryotes. Because of the simplicity of bacteria relative to larger organisms and the ease with which they can be manipulated experimentally, the cell structure of bacteria has been well studied, revealing many biochemical principles that have been subsequently applied to other organisms.

Cell Morphology

Perhaps the most elemental structural property of bacteria is cell morphology (shape). Typical examples include:

- coccus (spherical)
- bacillus (rod-like)
- spirillum (spiral)
- filamentous

Cell shape is generally characteristic of a given bacterial species, but can vary depending on growth conditions. Some bacteria have complex life cycles involving the production of stalks and appendages (e.g. *Caulobacter*) and some produce elaborate structures bearing reproductive spores (e.g. *Myxococcus*, *Streptomyces*).

Bacteria generally form distinctive cell morphologies when examined by light microscopy and distinct colony morphologies when grown on Petri plates. These are often the first characteristics observed by a microbiologist to determine the identity of an unknown bacterial culture.

Cell Wall

Perhaps the most obvious structural characteristic of bacteria is (with some exceptions) their small size. For example, *Escherichia coli* cells, an "average" sized bacterium, are about 2 micrometres (ìm) long and 0.5 ìm in diameter, with a cell volume of 0.6 - 0.7 ìm^3. This corresponds to a wet mass of about 1 picogram (pg), assuming that the cell consists mostly of water. The dry mass of a single cell can be estimated as 20% of the wet mass, amounting to 0.2 pg. About half of the dry mass of a bacterial cell consists of carbon, and also about half of it can be attributed to proteins. Therefore, a typical fully grown 1-liter culture of *Escherichia coli* (at an optical density of 1.0, corresponding to ca. 10^9 cells/ml) yields about 1 g wet cell mass.

Small size is extremely important because it allows for a large surface area-to-volume ratio which allows for rapid uptake and intracellular distribution of nutrients and excretion of wastes. At low surface area-to-volume ratios the diffusion of nutrients and waste products across the bacterial cell membrane limits the rate at which microbial metabolism can occur, making the cell less evolutionarily fit. The reason for the existence of large cells is unknown, although it is speculated that the increased cell volume is used primarily for storage of excess nutrients.

The cell envelope is composed of the plasma membrane and cell wall. As in other organisms, the bacterial cell wall provides structural integrity to the cell. In prokaryotes, the primary function of the cell wall is to protect the cell from internal turgor pressure caused by the much higher concentrations of proteins and other molecules inside the cell compared to its external environment.

The bacterial cell wall differs from that of all other organisms by the presence of peptidoglycan (poly-*N*-acetylglucosamine and *N*-acetylmuramic acid), which is located immediately outside of the cytoplasmic membrane. Peptidoglycan is responsible for the rigidity of the bacterial cell wall and for the determination of cell shape. It is relatively porous and is not considered to be a permeability barrier for small substrates.

While all bacterial cell walls (with a few exceptions e.g. extracellular parasites such as *Mycoplasma*) contain peptidoglycan, not all cell walls have the same overall structures. Since the cell wall is required for bacterial survival, but is absent in eukaryotes, several antibiotics (penicillins and cephalosporins) stop bacterial infections by interfering with cell wall synthesis, while having no effects on human cells, because human cells don't have cell walls, they only have cell membranes. There are two main types of bacterial cell walls, Gram positive and Gram negative, which are differentiated by their Gram staining characteristics. For both Gram-positive and Gram-negative bacteria, particles of approximately 2 nm can pass through the peptidoglycan. If the bacterial cell wall is entirely removed, it is called a protoplast while if it's partially removed, it is called a spheroplast. â-Lactam antibiotics such as penicillin inhibit the formation of peptidoglycan cross-links in the bacterial cell wall. The enzyme lysozyme, found in human tears, also digests the cell wall of bacteria and is the body's main defence against eye infections.

The Gram Positive Cell Wall

Gram positive cell walls are thick and peptidoglycans (mucopeptides, glycopeptides, mureins), the structural elements of almost all bacterial cell walls, constitute almost 95% of the cell wall in some Gram positive bacteria and as little as 5-10% of the cell wall in Gram negative bacteria. Peptidoglycans are made up of a polysaccharide backbone consisting of alternating N-acetylmuramic acid(NAM) and N-acetylglucosamine(NAG) residues in equal amounts. The cell wall of some Gram positive bacteria can be completely dissolved by lysozyme, as this enzyme attacks the bonds between GA and MA. In other Gram positive bacteria, e.g. Staphylococcus aureus, the walls are resistant to the action of lysozyme. They have O-acetyl groups on carbon-6 of some MA residues. The matrix substances in the walls of Gram positive bacteria may be polysaccharides or teichoic acids. The latter are very widespread, but have been found only in Gram positive bacteria. There are two main types of teichoic acid: ribitol teichoic acids and glycerol teichoic acids. The latter one is more widespread. These acids are polymers of ribitol phosphate and glycerol phosphate, respectively, and only located on the surface of many Gram positive bacteria. However, the exact function of teichoic acid is debated and not fully understood.

The Gram Negative Cell Wall

Gram negative cell walls are thin and unlike the Gram positive cell walls, they contain a thin peptidoglycan layer adjacent to the cytoplasmic membrane. This is responsible for the cell wall's inability to retain the crystal violet stain upon decolourisation (differentiation) with ethanol-acetic acid during Gram staining. In addition to the peptidoglycan layer, the Gram negative cell wall also contains an outer membrane composed by phospholipids and lipopolysaccharides, which face into the external environment. As the lipopolysaccharides are highly charged, the Gram negative cell wall has an overall negative charge. The chemical structure of the outer membrane lipopolysaccharides is often unique to specific bacterial strains (i.e. sub-species) and is responsible for many of the antigenic properties of these strains. Gram negative cell walls consist of a few layers of peptidoglycan surrounded by a second lipid membrane containing lipopolysaccharides and lipoproteins. Lipopolysaccharides, also called endotoxins, are composed of polysaccharides and lipid A (responsible for much of the toxicity of Gram-negative bacteria).

Plasma Membrane

The plasma membrane or bacterial cytoplasmic membrane is composed of a phospholipid bilayer and thus has all of the general functions of a cell membrane such as acting as a permeability barrier for most molecules and serving as the location for the transport of molecules into the cell. In addition to these functions, prokaryotic membranes also function in energy conservation as the location about which a proton motive force is generated. Unlike eukaryotes, bacterial membranes (with some exceptions e.g. *Mycoplasma* and methanotrophs) generally do not contain sterols.

However, many microbes do contain structurally related compounds called hopanoids which likely fulfill the same function. Unlike eukaryotes, bacteria can have a wide variety of fatty acids within their membranes. Along with typical saturated and unsaturated fatty acids, bacteria can contain fatty acids with additional methyl, hydroxy or even cyclic groups. The relative proportions of these fatty acids can be modulated by the bacterium to maintain the optimum fluidity of the membrane (e.g. following temperature change).

As a phospholipid bilayer, the lipid portion of the outer membrane is impermeable to charged molecules. However, channels called porins are present in the outer membrane that allow for passive transport of many ions, sugars and amino acids across the outer membrane. These molecules are therefore present in the periplasm, the region between the cytoplasmic and outer membranes. The periplasm contains the peptidoglycan layer and many proteins responsible for substrate binding or hydrolysis and reception of extracellular signals. The periplasm is thought to exist in a gel-like state rather than a liquid due to the high concentration of proteins and peptidoglycan found within it. Because of its location between the cytoplasmic and outer membranes, signals received and substrates bound are available to be transported across the cytoplasmic membrane using transport and signalling proteins imbedded there.

Extracellular (External) Structures

Fimbrae and Pili: Fimbrae (sometimes called "attachment pili") are protein tubes that extend out from the outer membrane in many members of the *Proteobacteria*. They are generally short in length and present in high numbers about the entire bacterial cell surface. Fimbrae usually function to facilitate the attachment of a bacterium to a surface (e.g. to form a biofilm) or to other cells (e.g. animal cells during pathogenesis).

A few organisms (e.g. *Myxococcus*) use fimbrae for motility to facilitate the assembly of multicellular structures such as fruiting bodies. Pili are similar in structure to fimbrae but are much longer and present on the bacterial cell in low numbers. Pili are involved in the process of bacterial conjugation where they are called conjugation pili or "sex pili". Type IV pili (non-sex pili) also aid bacteria in gripping surfaces.

S-layers

An S-layer (surface layer) is a cell surface protein layer found in many different bacteria and in some archaea, where it serves as the cell wall. All S-layers are made up of a two-dimensional array of proteins and have a crystalline appearance, the symmetry of which differs between species.

The exact function of S-layers is unknown, but it has been suggested that they act as a partial permeability barrier for large substrates. For example, an S-layer could conceivably keep extracellular proteins near the cell membrane by preventing their diffusion away from the cell. In some pathogenic species, an S-layer may help to facilitate survival within the host by conferring protection against host defence mechanisms.

Glycocalyx

Many bacteria secrete extracellular polymers outside of their cell walls called glycocalyx. These polymers are usually composed of polysaccharides and sometimes protein. Capsules are relatively impermeable structures that cannot be stained with dyes such as India ink. They are structures that help protect bacteria from phagocytosis and desiccation. Slime layer is involved in attachment of bacteria to other cells or inanimate surfaces to form biofilms. Slime layers can also be used as a food reserve for the cell.

Flagella

Perhaps the most recognizable extracellular bacterial cell structures are flagella. Flagella are whip-like structures protruding from the bacterial cell wall and are responsible for bacterial motility (i.e. movement). The arrangement of flagella about the bacterial cell is unique to the species observed. Common forms include:

- Monotrichous - Single flagellum
- Lophotrichous - A tuft of flagella found at one of the cell pole
- Amphitrichous - Single flagellum found at each of two opposite poles
- Peritrichous - Multiple flagella found at several locations about the cell

The bacterial flagellum consists of three basic components: a whip-like filament, a motor complex, and a hook that connects them. The filament is approximately 20 nm in diameter and consists of several protofilaments, each made up of thousands of flagellin subunits. The bundle is held together by a cap and may or may not be encapsulated. The motor complex consists of a series of rings anchoring the flagellum in the inner and outer membranes, followed by a proton-driven motor that drives rotational movement in the filament.

Intracellular (Intenal) Structures

In comparison to eukaryotes, the intracellular features of the bacterial cell are extremely simple. Bacteria do not contain organelles in the same sense as eukaryotes. Instead, the chromosome and perhaps ribosomes are the only easily observable intracellular structures found in all bacteria. They do exist, however, specialized groups of bacteria that contain more complex intracellular structures, some of which are discussed below.

The Bacterial Chromosome and Plasmids

Unlike eukaryotes, the bacterial chromosome is not enclosed inside of a membrane-bound nucleus but instead resides inside the bacterial cytoplasm. This means that the transfer of cellular information through the processes of translation, transcription and DNA replication all occur within the same compartment and can interact with other cytoplasmic structures, most notably ribosomes. The bacterial chromosome is not packaged using histones to form chromatin as in eukaryotes but instead exists as a highly compact supercoiled structure, the precise nature of which remains unclear. Most bacterial chromosomes are circular although some examples of linear chromosomes exist (e.g. *Borrelia burgdorferi*). Along with chromosomal DNA, most bacteria also contain small independent pieces of DNA called plasmids that often encode for traits that are advantageous but not essential to their bacterial host. Plasmids can be easily gained or lost by a bacterium and can be transferred between bacteria as a form of horizontal gene transfer.

Ribosomes and Other Multiprotein Complexes

In most bacteria the most numerous intracellular structure is the ribosome, the site of protein synthesis in all living organisms. All prokaryotes have 70S (where S=Svedberg units) ribosomes while eukaryotes contain larger 80S ribosomes in their cytosol. The 70S ribosome is made up of a 50S and 30S subunits. The 50S subunit contains the 23S and 5S rRNA while the 30S subunit contains the 16S rRNA. These rRNA molecules differ in size in eukaryotes and are complexed with a large number of ribosomal proteins, the number and type of which can vary slightly between organisms. While the ribosome is the most commonly observed intracellular multiprotein complex in bacteria other large complexes do occur and can sometimes be seen using microscopy.

Intracellular Membranes

While not typical of all bacteria some microbes contain intracellular membranes in addition to (or as extensions of) their cytoplasmic membranes. An early idea was that bacteria might contain membrane folds termed mesosomes, but these were later shown to be artifacts produced by the chemicals used to prepare the cells for electron microscopy. Examples of bacteria containing intracellular membranes are phototrophs, nitrifying bacteria and methane-oxidising bacteria. Intracellular membranes are also found in bacteria belonging to the poorly studied Planctomycetes group, although these membranes more closely resemble organellar membranes in eukaryotes and are currently of unknown function. Chromatophores are intracellular membranes found in phototrophic bacteria. Used primarily for photosynthesis, they contain bacteriochlorophyll pigments and carotenoids.

Cytoskeleton

The prokaryotic cytoskeleton is the collective name for all structural filaments in prokaryotes. It was once thought that prokaryotic cells did not possess cytoskeletons, but recent advances in visualization technology and structure determination have shown that filaments indeed exist in these cells. In fact, homologues for all major cytoskeletal proteins in eukaryotes have been found in prokaryotes. Cytoskeletal elements play essential roles in cell division, protection, shape determination, and polarity determination in various prokaryotes.

Nutrient Storage Structures

Most bacteria do not live in environments that contain large amounts of nutrients at all times. To accommodate these transient levels of nutrients bacteria contain several different methods of nutrient storage in times of plenty for use in times of want. For example, many bacteria store excess carbon in the form of polyhydroxyalkanoates or glycogen. Some microbes store soluble nutrients such as nitrate in vacuoles. Sulphur is most often stored as elemental (S^0) granules which can be deposited either intra- or extracellularly. Sulphur granules are especially common in bacteria that use hydrogen sulfide as an electron source. Most of the above mentioned examples can be viewed using a microscope and are surrounded by a thin nonunit membrane to separate them from the cytoplasm.

Inclusions

Inclusions are considered to be nonliving components of the cell that do not possess metabolic activity and are not bounded by membranes. The most common inclusions are glycogen, lipid droplets, crystals, and pigments. Volutin granules are cytoplasmic inclusions of complexed inorganic polyphosphate. These granules are called metachromatic granules due to their displaying the metachromatic effect; they appear red or blue when stained with the blue dyes methylene blue or toluidine blue.

Gas Vacuoles

Gas vacuoles are membrane-bound, spindle-shaped vesicles, found in some planktonic bacteria and *Cyanobacteria*, that provides buoyancy to these cells by decreasing their overall cell density. Positive buoyancy is needed to keep the cells in the upper reaches of the water column, so that they can continue to perform photosynthesis. They are made up of a shell of protein that has a highly hydrophobic inner surface, making it impermeable to water (and stopping water vapour from condensing inside) but permeable to most gases. Because the gas vesicle is a hollow cylinder, it is liable to collapse when the surrounding pressure becomes too great. Natural selection has fine tuned the structure of the gas vesicle to maximise its resistance to buckling, including an external strengthening protein, GvpC, rather like the green thread in a braided hosepipe. There is a simple relationship between the diameter of the gas vesicle and pressure at which it will collapse - the wider the gas vesicle the weaker it becomes.

However, wider gas vesicles are more efficient, providing more buoyancy per unit of protein than narrow gas vesicles. Different species produce gas vesicle of different diameter, allowing them to colonise different depths of the water column (fast growing, highly competitive species with wide gas vesicles in the top most layers; slow growing, dark-adapted, species with strong narrow gas vesicles in the deeper layers).

The diameter of the gas vesicle will also help determine which species survive in different bodies of water. Deep lakes that experience winter mixing, will exposes the cells to the hydrostatic pressure generated by the full water column. This will select for species with narrower, stronger gas vesicles.

The cell achieves its height in the water column by synthesising gas vesicles. As the cell rises up, it is able to increase its carbohydrate load through increased photosynthesis. Too high and the cell will suffer photobleaching and possible death, however, the carbohydrate produced during photosynthesis increases the cell's density, causing it to sink.

The daily cycle of carbohydrate build-up from photosynthesis and carbohydrate catabolism during dark hours is enough to fine tune the cell's position in the water column, bring it up toward the surface when its carbohydrate levels are low and it needs to photosynthesis, and allowing it to sink away from the harmful UV radiation when the cell's carbohydrate levels have been replenished. An extreme excess of carbohydrate causes a significant change in the internal pressure of the cell, which causes the gas vesicles to buckle and collapse and the cell to sink out.

Microcompartments

Bacterial microcompartments are widespread, membrane-bound organelles that are made of a protein shell that surrounds and encloses various enzymes. provide a further level of organization; they are compartments within bacteria that are surrounded by polyhedral protein shells, rather than by lipid membranes. These "polyhedral organelles" localize and compartmentalize bacterial metabolism, a function performed by the membrane-bound organelles in eukaryotes.

Carboxysomes

Carboxysomes are bacterial microcompartments found in many autotrophic bacteria such as Cyanobacteria, Knallgasbacteria, Nitroso- and Nitrobacteria. They are proteinaceous structures resembling phage heads in their morphology and contain the enzymes of carbon dioxide fixation in these organisms (especially ribulose bisphosphate carboxylase/oxygenase, RuBisCO, and carbonic anhydrase).

It is thought that the high local concentration of the enzymes along with the fast conversion of bicarbonate to carbon dioxide by carbonic anhydrase allows faster and more efficient carbon dioxide fixation than possible inside the cytoplasm. Similar structures are known to harbor the coenzyme B12-containing glycerol dehydratase, the key enzyme of glycerol fermentation to 1,3-propanediol, in some Enterobacteriaceae (e. g. Salmonella).

Magnetosomes

Magnetosomes are bacterial microcompartments found in magnetotactic bacteria that allow them to sense and align themselves along a magnetic field (magnetotaxis). The ecological role of magnetotaxis is unknown but it is hypothesized to be involved in the determination of optimal oxygen concentrations. Magnetosomes are composed of the mineral magnetite or greigite and are surrounded by a lipid bilayer membrane. The morphology of magnetosomes is species-specific.

Endospores

Perhaps, the most well known bacterial adaptation to stress is the formation of endospores. Endospores are bacterial survival structures that are highly resistant to many different types of chemical and environmental stresses and therefore enable the survival of bacteria in environments that would be lethal for these cells in their normal vegetative form.

It has been proposed that endospore formation has allowed for the survival of some bacteria for hundreds of millions of years (e.g. in salt crystals) although these publications have been questioned. Endospore formation is limited to several genera of Gram-positive bacteria such as *Bacillus* and *Clostridium*. It differs from reproductive spores in that only one spore is formed per cell resulting in no net gain in cell number upon endospore germination.

The location of an endospore within a cell is species-specific and can be used to determine the identity of a bacterium. Dipicolinic acid is a chemical compound which composes 5% to 15% of the dry weight of bacterial spores. It is implicated as responsible for the heat resistance of the endospore.

Cytoskeleton

The cytoskeleton (also CSK) is a cellular scaffolding or skeleton contained within a cell's cytoplasm. The cytoskeleton is present in all cells; it was once thought to be unique to eukaryotes, but recent research has identified the prokaryotic cytoskeleton. It forms structures such as flagella, cilia and lamellipodia and plays important roles in both intracellular transport (the movement of vesicles and organelles, for example) and cellular division.

In 1903 Nikolai K Koltsov proposed that the shape of cells was determined by a network of tubules which he termed the cytoskeleton. The concept of a protein mosaic that dynamically coordinated cytoplasmic biochemistry was proposed by Rudolph Peters in 1929 while the term (*cytosquelette*, in French) was first introduced by French embryologist Paul Wintrebert in 1931.

The Eukaryotic Cytoskeleton

Eukaryotic cells contain three main kinds of cytoskeletal filaments, which are microfilaments, intermediate filaments, and microtubules. The cytoskeleton provides the cell with structure and shape, and by excluding macromolecules from some of the cytosol it adds to the level of macromolecular crowding in this compartment.

Cytoskeletal elements interact extensively and intimately with cellular membranes. A number of small molecule cytoskeletal drugs have been discovered that interact with actin and microtubules. These compounds have proven useful in studying the cytoskeleton and several have clinical applications.

Microfilaments (Actin Filaments)

These are the thinnest filaments of the cytoskeleton. They are composed of linear polymers of actin subunits, and generate force by elongation at one end of the filament coupled with shrinkage at the other, causing net movement of the intervening strand. They also act as tracks for the movement of myosin molecules that attach to the microfilament and "walk" along them.

Actin structures are controlled by the Rho family of small GTP-binding proteins such as Rho itself for contractile acto-myosin filaments ("stress fibres"), Rac for lamellipodia and Cdc42 for filopodia.

Intermediate Filaments

These filaments, averaging 10 nanometers in diameter, are more stable (strongly bound) than actin filaments, and heterogeneous constituents of the cytoskeleton. Like actin filaments, they function in the maintenance of cell-shape by bearing tension (microtubules, by contrast, resist compression. It may be useful to think of micro- and intermediate filaments as cables, and of microtubules as cellular support beams).

Intermediate filaments organize the internal tridimensional structure of the cell, anchoring organelles and serving as structural components of the nuclear lamina and sarcomeres. They also participate in some cell-cell and cell-matrix junctions.

Different intermediate filaments are:

- made of vimentins, being the common structural support of many cells.
- made of keratin, found in skin cells, hair and nails.
- neurofilaments of neural cells.
- made of lamin, giving structural support to the nuclear envelope.

Microtubules

Microtubules are hollow cylinders about 23 nm in diameter (lumen = approximately 15 nm in diameter), most commonly comprising 13 protofilaments which, in turn, are polymers of alpha and beta tubulin. They have a very dynamic behaviour, binding GTP for polymerization. They are commonly organized by the centrosome.

In nine triplet sets (star-shaped), they form the centrioles, and in nine doublets oriented about two additional microtubules (wheel-shaped) they form cilia and flagella. The latter formation is commonly referred to as a "9+2" arrangement, where in each doublet is connected to another by the protein dynein. As both flagella and cilia are structural components of the cell, and are maintained by microtubules, they can be considered part of the cytoskeleton.

They play key roles in:

- intracellular transport (associated with dyneins and kinesins, they transport organelles like mitochondria or vesicles).
- the axoneme of cilia and flagella.
- the mitotic spindle.
- synthesis of the cell wall in plants.

Comparison

Cytoskeleton type	*Diameter (nm)*	*Structure*	*Subunit examples*
Microfilaments	6	double helix	actin
Intermediate filaments	10	two anti-parallel helices/dimers, forming tetramers	• vimentin (mesenchyme) • glial fibrillary acidic protein (glial cells) • neurofilament proteins (neuronal processes) • keratins (epithelial cells) • nuclear lamins
Microtubules	23	protofilaments, in turn consisting of tubulin subunits in complex with stathmin	α- and β-tubulin

The Prokaryotic Cytoskeleton

The cytoskeleton was previously thought to be a feature only of eukaryotic cells, but homologues to all the major proteins of the eukaryotic cytoskeleton have recently been found in prokaryotes. Although the evolutionary relationships are so distant that they are not obvious from protein sequence comparisons alone, the similarity of their three-dimensional structures and similar functions in maintaining cell shape and polarity provides strong evidence that the eukaryotic and prokaryotic cytoskeletons are truly homologous. However, some structures in the bacterial cytoskeleton may have yet to be identified.

FtsZ

FtsZ was the first protein of the prokaryotic cytoskeleton to be identified. Like tubulin, FtsZ forms filaments in the presence of GTP, but these filaments do not group into tubules. During cell division, FtsZ is the first protein to move to the division site, and is essential for recruiting other proteins that synthesize the new cell wall between the dividing cells.

MreB and ParM

Prokaryotic actin-like proteins, such as MreB, are involved in the maintenance of cell shape. All non-spherical bacteria have genes encoding actin-like proteins, and these proteins form a helical network beneath the cell membrane that guides the proteins involved in cell wall biosynthesis.

Some plasmids encode a partitioning system that involves an actin-like protein ParM. Filaments of ParM exhibit dynamic instability, and may partition plasmid DNA into the dividing daughter cells by a mechanism analogous to that used by microtubules during eukaryotic mitosis.

Crescentin

The bacterium *Caulobacter crescentus* contains a 3rd protein, crescentin, that is related to the intermediate filaments of eukaryotic cells. Crescentin is also involved in maintaining cell shape, such as helical and vibrioid forms of bacteria, but the mechanism by which it does this is currently unclear.

History

Microtrabeculae: A fourth eukaryotic cytoskeletal element, *microtrabeculae*, was proposed by Keith Porter based on images obtained from high-voltage electron microscopy of whole cells in the 1970s. The images showed short, filamentous structures of unknown molecular composition associated with known cytoplasmic structures. Porter proposed that this microtrabecular structure represented a novel filamentous network distinct from microtubules, filamentous actin, or intermediate filaments.

It is now generally accepted that microtrabeculae are nothing more than an artifact of certain types of fixation treatment, although we have yet to fully understand the complexity of the cell's cytoskeleton.

Chytridiomycetes

Chytridiomycetes is a class of fungi. Members are found in soil, fresh water, and saline estuaries. They are primitive fungi, and are first known from the Rhynie chert. It has recently been redefined to exclude the taxa Neocallimastigomycota and Monoblepharidomycetes, which are now a phylum and a sister-class respectively. It includes the genera *Olpidiopsis* and *Hypochytrium* in addition to the type genus.

Chytridiomycetes is the major class of the phylum Chytridiomycota, which is responsible for the parasitic infections of organisms such as *Daphnia* and a number of amphibian species, among others.

Saccharomycetes

Saccharomycetes is a class in the kingdom of fungi. It contains the order Saccharomycetales, the budding yeasts. Hemiascomycetes is a more or less synonymous name.

Aspergillus

Aspergillus is a genus consisting of several hundred mold species found in various climates worldwide.

Aspergillus was first catalogued in 1729 by the Italian priest and biologist Pier Antonio Micheli. Viewing the fungi under a microscope, Micheli was reminded of the shape of an aspergillum (holy water sprinkler), from Latin *spargere* (to sprinkle), and named the genus accordingly. Today "aspergillum" is also the name of an asexual spore-forming structure common to all Aspergilli; around one-third of species are also known to have a sexual stage.

Growth and Distribution

Aspergillus species are highly aerobic and are found in almost all oxygen-rich environments, where they commonly grow as molds on the surface of a substrate, as a result of the high oxygen tension. Commonly, fungi grow on carbon-rich substrates like monosaccharides (such as glucose) and polysaccharides (such as amylose). *Aspergillus* species are common contaminants of starchy foods (such as bread and potatoes), and grow in or on many plants and trees.

In addition to growth on carbon sources, many species of *Aspergillus* demonstrate oligotrophy where they are capable of growing in nutrient-depleted environments, or environments in which there is a complete lack of key nutrients. *A. niger* is a prime example of this; it can be found growing on damp walls, as a major component of mildew.

Commercial Importance

Species of *Aspergillus* are important medically and commercially. Some species can cause infection in humans and other animals. Some infections found in animals have been studied for years. Some species found in animals have been described as new and specific to the investigated disease and others have been known as names already in use for organisms such as saprophytes. More than 60 *Aspergillus* species are medically relevant pathogens. For humans there is a range of diseases such as infection to the external ear, skin lesions, and ulcers classed as mycetomas.

Other species are important in commercial microbial fermentations. For example, alcoholic beverages such as Japanese sake are often made from rice or other starchy ingredients (like manioc), rather than from grapes or malted barley. Typical microorganisms used to make alcohol, such as yeasts of the genus

Saccharomyces, cannot ferment these starches, and so *koji* mold such as *Aspergillus oryzae* is used to break down the starches into simpler sugars.

Members of the genus are also sources of natural products that can be used in the development of medications to treat human disease.

Perhaps the largest application of *Aspergillus niger* is as the major source of citric acid; this organism accounts for over 99% of global citric acid production, or more than 1.4 million tonnes per annum. *A. niger* is also commonly used for the production of native and foreign enzymes, including glucose oxidase and hen egg white lysozyme. In these instances, the culture is rarely grown on a solid substrate, although this is still common practice in Japan, but is more often grown as a submerged culture in a bioreactor. In this way, the most important parameters can be strictly controlled, and maximal productivity can be achieved. It also makes it far easier to separate the chemical or enzyme of importance from the medium, and is therefore far more cost-effective.

Research

A. nidulans has been used as a research organism for many years and was used by Guido Pontecorvo to demonstrate parasexuality in fungi. Recently, *A. nidulans* was one of the pioneering organisms to have its genome sequenced by researchers at the Broad Institute. As of 2008, a further seven *Aspergillus* species have had their genomes sequenced: the industrially useful *A. niger* (two strains), *A. oryzae* and *A. terreus*, and the pathogens *A. clavatus*, *A. fischerianus* (*Neosartorya fischeri*), *A. flavus* and *A. fumigatus* (two strains). *A. fischerianus* is hardly ever pathogenic, but is very closely related to the common pathogen *A. fumigatus*; it was sequenced in part to better understand *A. fumigatus* pathogenicity.

Genomics

The simultaneous publication of three *Aspergillus* genome manuscripts in *Nature* in December 2005 established the genus as the leading filamentous fungal genus for comparative genomic studies. Like most major genome projects, these efforts were collaborations between a large sequencing centre and the respective community of scientists. For example, the Institute for Genome Research (TIGR) worked with the *A. fumigatus* community. *A. nidulans* was sequenced at the Broad Institute. *A. oryzae* was sequenced in Japan at the National Institute of Advanced Industrial Science and Technology.

The Joint Genome Institute of the Department of Energy has released sequence data for a citric acid-producing strain of *A. niger*. TIGR, now renamed the Venter Institute, is currently spearheading a project on the *A. flavus* genome.

Genome sizes for sequenced species of *Aspergillus* range from about 29.3 Mb for *A. fumigatus* to 37.1 Mb for *A. oryzae*, while the numbers of predicted genes vary from about 9926 for *A. fumigatus* to about 12,071 for *A. oryzae*. The genome size of an enzyme-producing strain of *A. niger* is of intermediate size at 33.9 Mb.

Pathogens

Some *Aspergillus* species cause serious disease in humans and animals. The most common pathogenic species are *A. fumigatus* and *A. flavus*, which produces aflatoxin which is both a toxin and a carcinogen, and which can contaminate foods such as nuts.

The most common species causing allergic disease are *A. fumigatus* and *Aspergillus clavatus*. Other species are important as agricultural pathogens. *Aspergillus* spp. cause disease on many grain crops, especially maize, and synthesize mycotoxins, including aflatoxin.

Aspergillosis

Aspergillosis is the group of diseases caused by *Aspergillus*. The most common subtype among paranasal sinus infections associated with aspergillosis is *A. fumigatus*.

The symptoms include fever, cough, chest pain, or breathlessness, which also occur in many other illnesses, so diagnosis can be difficult. Usually, only patients with already weakened immune systems or who suffer other lung conditions are susceptible.

In humans, the major forms of disease are:

- Allergic bronchopulmonary aspergillosis, which affects patients with respiratory diseases such as asthma, cystic fibrosis, and sinusitis
- Acute invasive aspergillosis, a form that grows into surrounding tissue, more common in those with weakened immune systems such as AIDS or chemotherapy patients
- Disseminated invasive aspergillosis, an infection spread widely through the body
- Aspergilloma, a "fungus ball" that can form within cavities such as the lung

Aspergillosis of the air passages is also frequently reported in birds, and certain species of *Aspergillus* have been known to infect insects.

Heterothallic

Heterothallic species have sexes that reside in different individuals. The term is applied particularly to distinguish heterothallic fungi, which require two compatible partners to produce sexual spores, from homothallic ones, which are capable of sexual reproduction from a single organism.

Pezizales

The Pezizales are an order of the subphylum Pezizomycotina within the phylum Ascomycota. The order contains 16 families, 199 genera, and 1683 species. It contains a number of species of economic importance, such as morels, the black and white truffles, and the desert truffles. The Pezizales are saprobic, mycorrhizal, or parasitic on plants. Species grow on soil, wood, leaves and dung. Soil-inhabiting species often fruit in habitats with a high pH and low content of organic matter, including disturbed ground. Most species occur in temperate regions or at high elevation. Several members of the Sarcoscyphaceae and Sarcosomataceae are common in tropical regions.

Description

Members of this order are characterized by asci that typically open by rupturing to form a terminal or eccentric lid or operculum. The ascomata are apothecia or are closed structures of various forms derived from apothecia. Apothecia range in size from less than a millimeter to approximately 15 cm, and may be stalked or sessile. The order includes epigeous, semihypogeous to hypogeous (truffles) taxa. The ascospores are single-celled, bipolar symmetrical, and usually bilaterally symmetrical, ranging from roughly spherical to ellipsoidal to occasionally fusoid. The ascospores of some species develop surface ornamentations such as warts, ridges, or spines. The tissues of the ascomata are fleshy and often fragile. Although the majority of species are known only in the teleomorphic state, the anamorphs of some species are known.

Peronosporales

The Peronosporales are an order of water moulds (Class Oomycetes) which can be pathogenic. Many diseases of plants are sometimes classified under this order, but are sometimes considered members of order Pythiales. Some of these pathogenic protists include the organisms responsible for potato blight, eucalyptus dieback, sudden oak death, and blue mold. Further genetic studies may place these organisms more definitively in one order or another.

Lichen

Lichens are composite organisms consisting of a fungus (the mycobiont) and a photosynthetic partner (the photobiont or phycobiont) growing together in a symbiotic relationship. The photobiont is usually either a green alga (commonly *Trebouxia*) or cyanobacterium (commonly *Nostoc*). The morphology, physiology and biochemistry of lichens are very different from those of the isolated fungus and alga in culture. Lichens occur in some of the most extreme environments on Earth—arctic tundra, hot deserts, rocky coasts, and toxic slag heaps. However, they are also abundant as epiphytes on leaves and branches in rain forests and temperate woodland, on bare rock, including walls and gravestones, and on exposed soil surfaces (e.g., *Collema*) in otherwise mesic habitats. The roofs of many buildings have lichens growing on them. Lichens are widespread and may be long-lived; however, many are also vulnerable to environmental disturbance, and may be useful to scientists in assessing the effects of air pollution, ozone depletion, and metal contamination. Lichens have also been used in making dyes and perfumes, as well as in traditional medicines.

Overview

The body (thallus) of most lichens is different from those of either the fungus or alga growing separately. The fungus surrounds the algal cells, often enclosing them within complex fungal tissues unique to lichen associations. In many species the fungus penetrates the algal cell wall, forming penetration pegs or haustoria similar to those produced by pathogenic fungi. Lichens are poikilohydric, capable of surviving extremely low levels of water content. However, the reconfiguration of membranes following a period of dehydration requires several minutes at least.

The algal or cyanobacterial cells are photosynthetic, and as in plants they reduce atmospheric carbon dioxide into organic carbon sugars to feed both symbionts. Both partners gain water and mineral nutrients mainly from the atmosphere, through rain and dust. The fungal partner protects the alga by retaining water, serving as a larger capture area for mineral nutrients and, in some cases, provides minerals obtained from the substrate. If a cyanobacterium is present, as a primary partner or another symbiont in addition to green alga as in certain tripartite lichens, they can fix atmospheric nitrogen, complementing the activities of the green alga.

Although strains of cyanobacteria found in various cyanolichens are often closely related to one another, they differ from the most closely related free-living strains. The lichen association is a close symbiosis. It extends the ecological range of both partners but is not always obligatory for their growth and reproduction in natural environments, since many of the algal symbionts can live independently. A prominent example is the alga *Trentepohlia* which forms orange-coloured populations on tree trunks and suitable rock faces. Lichen propagules (diaspores) typically contain cells from both partners, although the fungal components of so-called "fringe species" rely instead on algal cells dispersed by the "core species."

Lichen associations may be examples of mutualism, commensalism or even parasitism, depending on the species. Cyanobacteria in laboratory settings can grow faster when they are alone rather than when they are part of a lichen.

Recently, on 26 April 2012, scientists reported that lichen survived and showed remarkable results on the adaptation capacity of photosynthetic activity within the simulation time of 34 days under Martian conditions in the Mars Simulation Laboratory (MSL) maintained by the German Aerospace Centre (DLR).

History

Although lichens had been recognized as organisms for quite some time, it was not until 1867, when Swiss botanist Simon Schwendener proposed his dual theory of lichens, that the true nature of the lichen association began to emerge. Schwendener's hypothesis, which at the time lacked experimental evidence, arose from his extensive analysis of the anatomy and development in lichens, algae, and fungi using a light microscope. Many of the leading lichenologists at the time, such as James Crombie and Nylander, rejected Schwendener's hypothesis because the common consensus was that all living organisms were autonomous. Other prominent biologists, such as Heinrich Anton de Bary, Albert Bernhard Frank, Melchior Treub and Hermann Hellriegel were not so quick to reject Schwendener's ideas and the concept soon spread into other areas of study, such as microbial, plant, animal and human pathogens. When the complex relationships between pathogenic microorganisms and their hosts were finally identified—refuting the idea of holistic organisms—Schwendener's hypothesis began to gain popularity. Further experimental proof of the dual nature of lichens was obtained when Eugen Thomas published his results in 1939 on the first successful re-synthesis experiment.

Symbionts

Living as a symbiont in a lichen appears to be a very successful way for a fungus to derive essential nutrients, as about 20% of all fungal species have acquired this mode of life. The largest number of lichenized fungi occur in the Ascomycota, with about 40% of species forming such an association. Some of these lichenized fungi occur in orders with nonlichenized fungi that live as saprotrophs or plant parasites (for example, the Leotiales, Dothideales, and Pezizales). Other lichen fungi occur in only five orders in which all members are engaged in this habit (Orders Graphidales, Gyalectales, Peltigerales, Pertusariales, and Teloschistales). Lichenized and nonlichenized fungi can even be found in the same genus or species. Overall, about 98% of lichens have an ascomycetous mycobiont. Next to the Ascomycota, the largest number of lichenized fungi occur in the unassigned fungi imperfecti. Comparatively few Basidiomycetes are lichenized, but these include agarics, such as species of *Lichenomphalia*, clavarioid fungi, such as species of *Multiclavula*, and corticioid fungi, such as species of *Dictyonema*.

The autotrophic symbionts occurring in lichens are simple, photosynthetic organisms commonly and traditionally known as algae. These symbionts include both prokaryotic and eukaryotic organisms. Approximately 100 species of photosynthetic partners from 40 genera and five distinct classes (prokaryotic: Cyanophyceae; eukaryotic: Trebouxiophyceae, Phaeophyceae, Chlorophyceae) have been found to associate with the lichen-forming fungi. The prokaryotes belong to the Cyanobacteria, whose representatives are often called bluegreen algae. The bluegreen algae occur as symbionts in about 8% of the known lichens. The most commonly occurring genus is *Nostoc*. The majority of the lichens contain eukaryotic autotrophs belonging to the Chlorophyta (green algae) or to the Xanthophyta (yellow-green algae). About 90% of all known lichens have a green alga as a symbiont, and among these, *Trebouxia* is the most common genus, occurring in about 40% of all lichens. The second most commonly represented green alga genus is *Trentepohlia*. Overall, about 100 species are known to occur as autotrophs in lichens. All the algae are probably able to exist independently in nature as well as in the lichen.

A particular fungus species and algal species are not necessarily always associated together in a lichen. One fungus, for example, can form lichens with a variety of different algae. The thalli produced by a given fungal symbiont with its differing partners will be similar, and the secondary metabolites identical, indicating that the fungus has the dominant role in determining the morphology of the lichen. Further, the same algal species can occur in association with different fungal partners. Lichens are known in which there is one fungus associated with two or even three algal species. Rarely, the reverse can occur, and two or more fungal species can interact to form the same lichen.

Both the lichen and the fungus partner bear the same scientific name, and the lichens are being integrated into the classification schemes for fungi. The alga bears its own scientific name, which bears no relationship to that of the lichen or fungi.

Morphology and Structure

Some lichens have the aspect of leaves (foliose lichens); others cover the substrate like a crust (crustose lichens), others such as the genus *Ramalina* adopt shrubby forms (fruticose lichens), and there are gelatinous lichens such as the genus *Collema*.

Although the form of a lichen is determined by the genetic material of the fungal partner, association with a photobiont is required for the development of that form. When grown in the laboratory in the absence of its photobiont, a lichen fungus develops as an undifferentiated mass of hyphae. If combined with its photobiont under appropriate conditions, its characteristic form emerges, in the process called morphogenesis (Brodo, Sharnoff & Sharnoff, 2001). In a few remarkable cases, a single lichen fungus can develop into two very different lichen forms when associating with either a green algal or a cyanobacterial symbiont. Quite naturally, these alternative forms were at first considered to be different species, until they were first found growing in a conjoined manner.

There is evidence to suggest that the lichen symbiosis is parasitic or commensalistic, rather than mutualistic (Ahmadjian 1993). The photosynthetic partner can exist in nature independently of the fungal partner, but not vice versa. Furthermore, photobiont cells are routinely destroyed in the course of nutrient exchange. The association is able to continue because reproduction of the photobiont cells matches the rate at which they are destroyed. (ibid.)

Under magnification, a section through a typical foliose lichen thallus reveals four layers of interlaced fungal filaments. The uppermost layer is formed by densely agglutinated fungal hyphae building a protective outer layer called the cortex, which can reach several hundred ìm in thickness. This cortex may be further topped by an epicortex 0.6-1ìm thick in some Parmeliaceae, which may be with or without pores, and is secreted by cells—it is not itself cellular.

In lichens that include both green algal and cyanobacterial symbionts, the cyanobacteria may be held on the upper or lower surface in small pustules called cephalodia. Beneath the upper cortex is an algal layer composed of algal cells embedded in rather densely interwoven fungal hyphae. Each cell or group of cells of the photobiont is usually individually wrapped by hyphae, and in some cases penetrated by an haustorium. Beneath this algal layer is a third layer of loosely interwoven fungal hyphae without algal cells. This layer is called the medulla. Beneath the medulla, the bottom surface resembles the upper surface and is called the lower cortex, again consisting of densely packed fungal hyphae.

The lower cortex often bears rootlike fungal structures known as rhizines, which serve to attach the thallus to the substrate on which it grows. Lichens also sometimes contain structures made from fungal metabolites, for example crustose lichens sometimes have a polysaccharide layer in the cortex. Although each lichen thallus generally appears homogeneous, some evidence seems to suggest that the fungal component may consist of more than one genetic individual of that species. This seems to also be true of the photobiont species involved.

Growth Form

Lichens are informally classified by growth form into:

- crustose (paint-like, flat), e.g., *Caloplaca flavescens*
- filamentous (hair-like), e.g., *Ephebe lanata*
- foliose (leafy), e.g., *Hypogymnia physodes*
- fruticose (branched), e.g., *Cladonia evansii*, *C. subtenuis*, and *Usnea australis*
- leprose (powdery), e.g., *Lepraria incana*
- squamulose (consisting of small scale-like structures, lacking a lower cortex), e.g., *Normandina pulchella*
- gelatinous lichens, in which the cyanobacteria produce a polysaccharide that absorbs and retains water.

Reproduction and Dispersal

Many lichens reproduce asexually, either by vegetative reproduction or through the dispersal of diaspores containing algal and fungal cells. *Soredia* (singular soredium) are small groups of algal cells surrounded by fungal filaments that form in structures called *soralia*, from which the soredia can be dispersed by wind. Another form of diaspore are *isidia*, elongated outgrowths from the thallus that break off for mechanical dispersal. Fruticose lichens in particular can easily fragment. Because of the relative lack of differentiation in the thallus, the line between diaspore formation and vegetative reproduction is often blurred. Many lichens break up into fragments when they dry, dispersing themselves by wind action, to resume growth when moisture returns.

Many lichen fungi appear to reproduce sexually in a manner typical of fungi, producing spores that are presumably the result of sexual fusion and meiosis. Following dispersal, such fungal spores must meet with a compatible algal partner before a functional lichen can form. This may be a common form of reproduction in basidiolichens, which form fruitbodies resembling their nonlichenized relatives. Among the ascolichens, spores are produced in spore-producing bodies, the three most common spore body types are the apothecia, perithecia and the pycnidia.

For reproduction, lichen possess isidia, soredia, and undergo simple fragmentation. These structures are also composed of a fungal hyphae wrapped around cyanobacteria. (Eichorn, Evert, and Raven, 2005) While the reproductive structures are all composed of the same components (Mycobiont and Photobiont) they are each unique in other ways. Isidia are small outgrowths on the exterior of the lichen. Soredia are powdery propagules that are released from the top of the thallus. In order to establish the lichen, the soredia propagules must contain both the photobiont and the mycobiont.

Growth and Longevity

Lichenometry: Lichenometry is a technique used to determine the age of exposed rock surfaces based on the size of lichen thalli. Introduced by Beschel in the 1950s, the technique has found many applications.

Ecology

Lichens must compete with plants for access to sunlight, but because of their small size and slow growth, they thrive in places where higher plants have difficulty growing. Lichens are often the first to settle in places lacking soil, constituting the sole vegetation in some extreme environments such as those found at high mountain elevations and at high latitudes. Some survive in the tough conditions of deserts, and others on frozen soil of the Arctic regions.

A major ecophysiological advantage of lichens is that they are poikilohydric (*poikilo-* variable, *hydric-* relating to water), meaning that though they have little control over the status of their hydration, they can tolerate irregular and extended periods of severe desiccation. Like some mosses, liverworts, ferns, and a few "resurrection plants", upon desiccation, lichens enter a metabolic suspension or stasis (known as cryptobiosis) in which the cells of the lichen symbionts are dehydrated to a degree that halts most biochemical activity. In this cryptobiotic state, lichens can survive wider extremes of temperature, radiation and drought in the harsh environments they often inhabit.

Lichens do not have roots and do not need to tap continuous reservoirs of water like most higher plants, thus they can grow in locations impossible for most plants, such as bare rock, sterile soil or sand, and various artificial structures such as walls, roofs and monuments. Many lichens also grow as epiphytes (*epi-* on the surface, *phyte-* plant) on plants, particularly on the trunks and branches of trees. When growing on plants, lichens are not parasites; they do not consume any part of the plant nor poison it. Some ground-dwelling lichens, such as members of the subgenus *Cladina* (reindeer lichens), however, produce allelopathic chemicals which leach into the soil and inhibit the germination of plant seeds and growth of young plants. Stability (that is, longevity) of their substrate is a major factor of lichen habitats. Most lichens grow on stable rock surfaces or the bark of old trees, but many others grow on soil and sand. In these latter cases, lichens are often an important part of soil stabilization; indeed, in some desert ecosystems, vascular (higher) plant seeds cannot become established except in places where lichen crusts stabilize the sand and help retain water.

The European Space Agency has discovered that lichens can survive unprotected in space. In an experiment led by Leopoldo Sancho from the Complutense University of Madrid, two species of lichen—*Rhizocarpon geographicum* and *Xanthoria elegans*—were sealed in a capsule and launched on a Russian Soyuz rocket on 31 May 2005. Once in orbit the capsules were opened and the lichens were directly exposed to the vacuum of space with its widely fluctuating temperatures and cosmic radiation. After 15 days the lichens were brought back to earth and were found to be in full health with no discernible damage from their time in orbit.

When growing on mineral surfaces, some lichens slowly decompose their substrate by chemically degrading and physically disrupting the minerals, contributing to the process of weathering by which rocks are gradually turned into soil. While this contribution to weathering is usually benign, it can cause problems for artificial stone structures. For example, there is an ongoing lichen growth problem on Mount Rushmore National Memorial that requires the employment of mountain-climbing conservators to clean the monument.

Lichens may be eaten by some animals, such as reindeer, living in arctic regions. The larvae of a number of Lepidoptera species feed exclusively on lichens. These include Common Footman and Marbled Beauty. However, lichens are very low in protein and high in carbohydrates, making them unsuitable for some animals. Lichens are also used by the Northern Flying Squirrel for nesting, food, and a water source during winter.

Air Pollution

If lichens are exposed to air pollutants at all times, without any deciduous parts, they are unable to avoid the accumulation of pollutants. Also lacking stomata and a cuticle, lichens may absorb aerosols and gases over the entire thallus surface from which they may readily diffuse to the photobiont layer. Because lichens do not possess roots, their primary source of most elements is the air, and therefore elemental levels in lichens often reflect the accumulated composition of ambient air. The processes by which atmospheric deposition occurs include fog and dew, gaseous absorption, and dry deposition. Consequently, many environmental studies with lichens emphasize their feasibility as effective biomonitors of atmospheric quality.

Not all lichens are equally sensitive to air pollutants, so different lichen species show different levels of sensitivity to specific atmospheric pollutants. The sensitivity of a lichen to air pollution is directly related to the energy needs of the mycobiont, so that the stronger the dependency of the mycobiont on the photobiont, the more sensitive the lichen is to air pollution. Upon exposure to air pollution, the photobiont may use metabolic energy for repair of cellular structures that would otherwise be used for maintenance of photosynthetic activity, therefore leaving less metabolic energy available for the mycobiont. The alteration of the balance between the photobiont and mycobiont can lead to the breakdown of the symbiotic association. Therefore, lichen decline may result not only from the accumulation of toxic substances, but also from altered nutrient supplies that favour one symbiont over the other.

Evolution and Paleontology

The evolution of lichens and the phylum Ascomycota is complex and not well understood, but because there are fifteen different classes of Ascomycetes, scientists generally believe that different lichens have evolved independently from one another through analogous evolution. Lichenized fungi have continued to evolve, developing differently than those that do not form lichens.

Lichenization is an ancient nutritional strategy for fungi. The extreme habitats that lichens inhabit are not ordinarily conducive to producing fossils. The oldest fossil lichens in which both symbiotic partners have been recovered date to the Early Devonian Rhynie chert, about 400 million years old. The slightly older fossil *Spongiophyton* has also been interpreted as a lichen on morphological and isotopic grounds, although the isotopic basis is decidedly shaky. It has been suggested—although not yet proven—that the even older fossil *Nematothallus* was a lichen.

It has also been claimed that Ediacaran fossils were lichens, though this claim is controversial. Additional evidence has been marshalled for a lichen interpretation of *Dickinsonia*. Lichen-like fossils consisting of coccoid cells and thin filaments, preserved in marine phosphorite of the Doushantuo Formation in southern China. These fossils are thought to be 551 to 635 million years old (belonging to the Neoproterozoic era). Discovery of these fossils suggest that fungi developed symbiotic partnerships with photoautotrophs long before the evolution of vascular plants. *Winfrenatia*, an early zygomycetous lichen symbiosis that may have involved controlled parasitism, is an impression found in Scotland, belonging to the early Devonian times. There are also several examples of fossilized lichens embedded in amber. The fossilized *Anzia* is found in pieces of amber in northern Europe and dates back approximately 40 million years. Fossilized *Lobaria* comes from Trinity County in northern California, USA and dates back to the early to middle Miocene.

In 1995, Gargas and colleagues proposed that there were at least five independent origins of lichenization; three in the basidiomycetes and at least two in the Ascomycetes. However, Lutzoni et al. (2000)

indicate that lichenization probably evolved earlier and was followed by multiple independent losses. Some non-lichen-forming fungi may have secondarily lost the ability to form a lichen association. As a result, lichenization has been viewed as a highly successful nutritional strategy.

Lichens were a component of the early terrestrial ecosystems, and the estimated age of the oldest terrestrial lichen fossil is 400 Ma. Recent (2009) studies suggest that the ancestral ecological state of the Ascomycota was saprobism, and that independent lichenization events have occurred multiple times.

Taxonomy and Classification

Lichens are named based on the fungal component, which plays the primary role in determining the lichen's form. The fungus typically comprises the majority of a lichen's bulk, though in filamentous and gelatinous lichens this is not always the case. The lichen fungus is typically a member of the Ascomycota—rarely a member of the Basidiomycota, and then termed basidiolichens to differentiate them from the more common ascolichens.

Formerly, some lichen taxonomists placed lichens in their own division, the Mycophycophyta, but this practice is no longer accepted because the components belong to separate lineages. Neither the ascolichens nor the basidiolichens form monophyletic lineages in their respective fungal phyla, but they do form several major solely or primarily lichen-forming groups within each phylum. Even more unusual than basidiolichens is the fungus *Geosiphon pyriforme*, a member of the Glomeromycota that is unique in that it encloses a cyanobacterial symbiont inside its cells. *Geosiphon* is not usually considered to be a lichen, and its peculiar symbiosis was not recognized for many years. The genus is more closely allied to endomycorrhizal genera.

Human Use

Food: Lichens are eaten by many different cultures across the world. Although some lichens are only eaten in times of famine, others are a staple food or even a delicacy. Two obstacles are often encountered when eating lichens: lichen polysaccharides are generally indigestible to humans, and lichens usually contain mildly toxic secondary compounds that should be removed before eating. Very few lichens are poisonous, but those high in vulpinic acid or usnic acid are toxic. Most poisonous lichens are yellow.

In the past Iceland moss (*Cetraria islandica*) was an important human food in northern Europe, and was cooked as a bread, porridge, pudding, soup, or salad. Wila (*Bryoria fremontii*) was an important food in parts of North America, where it was usually pitcooked. Northern peoples in North America and Siberia traditionally eat the partially digested reindeer lichen (*Cladina* spp.) after they remove it from the rumen of caribou or reindeer that have been killed. Rock tripe (*Umbilicaria* spp. and *Lasalia* spp.) is a lichen that has frequently been used as an emergency food in North America, and one species, *Umbilicaria esculenta*, is used in a variety of traditional Korean and Japanese foods.

Other Uses

Many lichens produce secondary compounds, including pigments that reduce harmful amounts of sunlight and powerful toxins that reduce herbivory or kill bacteria. These compounds are very useful for lichen identification, and have had economic importance as dyes such as cudbear or primitive antibiotics.

There are reports dating almost 2000 years old of lichens being used to extract purple and red colours. Of great historical and commercial significance are lichens belonging to the family Roccellaceae, commonly called orchella weed or orchil. Orcein and other lichen dyes have largely been replaced by synthetic versions. The pH indicator litmus is a dye extracted from the lichen genus *Rocella tinctoria* by boiling.

Extracts from many *Usnea* species were used to treat wounds in Russia in the mid-twentieth century. Historically in Europe, *Lobaria pulmonaria* was collected in large quantities as "Lungwort" and, due to its lung-like appearance, was sold as a cure for lung diseases. Similarly *Peltigera leucophlebia* was used as a supposed cure for thrush, due to the resemblance of its cephalodia to the appearance of the disease.

The substance olivetol is found to be naturally present in certain species of lichens. This is a property it shares with the cannabis plant, which internally produces the related substance olivetolic acid (before using it to biosynthesise tetrahydrocannabinol (THC)).

Lichen is used in model railroading and other modelling hobbies as a material for making trees and shrubs.

An article published in *PLoS ONE* indicates that a serine protease isolated from certain species of lichen degrades prions.

Ethnolichenology

Ethnolichenology is the study of the relationship between lichens and people. Lichens have and are being used for many different purposes by human cultures on every continent, with the possible exception of Australia. The most common human use of lichens is for dye, but they are also frequently used for medicine and food. Different human cultures across the world have also found many other more novel uses for lichens.

Lichens for Dye

Lichens are a common source of natural dyes. The lichen dye is usually extracted by either boiling water or ammonia fermentation. Although usually called ammonia fermentation, this method is not actually a fermentation and involves letting the lichen steep in ammonia (traditionally urine) for at least two to three weeks.

In North America the most significant lichen dye is *Letharia vulpina*. Indigenous people through most of this lichen's range in North America traditionally make a yellow dye from this lichen by boiling it in water.

Historically, traditional dyes in Scotland were very important. Brown lichen dyes (called crottle) and red lichen dyes (called corkir) were used extensively to produce tartans. Purple dyes from lichens were historically very important throughout Europe from the 15th to 17th centuries. They were generally extracted from *Roccella* spp. lichens imported from the Canary Islands, Cape Verde Islands, Madagascar, or India. These lichens, and the dye extracted from them, are called orchil (variants archil, orchilla). The same dye was also produced from *Ochrolechia* spp. lichens in Britain and was called cudbear. Both *Roccella* spp. and *Ochrolechia* spp. contain the lichen substance orcin, which converts into the purple dye orcein in the ammonia fermentation process.

Litmus, a water soluble pH indicator dye mixture is also extracted from *Roccella* species.

Lichens for Medicine

Many lichens have been used medicinally across the world. A lichen's usefulness as a medicine is often related to the lichen secondary compounds that are abundant in most lichen thalli. Different lichens produce a wide variety of these compounds, most of which are unique to lichens and many of which are antibiotic. It has been estimated that 50% of all lichen species have antibiotic properties. Many lichen extracts have been found to be effective in killing Gram-positive bacteria, which included species that cause boils, scarlet fever, and pneumonia

One of the most potent lichen antibiotics is usnic acid, as a result *Usnea* spp. are commonly used in traditional medicines. *Usnea* was used in the United States as a fungal remedy of the mouth, stomach, intestines, anus, vagina, nose, ear, and skin, and in Finland it was used to treat wounds, skin eruptions, and athlete's foot. In Russia a preparation of the sodium salt of usnic acid was sold under the name Binan for the treatment of varicose and trophic ulcers, second and third degree burns, and for plastic surgery.

Other lichens commonly featured in folk medicines include Iceland moss and Lungwort.

Lichens for Poisons

Only a few lichens are truly poisonous, with species of *Letharia* and *Vulpicida* being the primary examples. These lichens are yellow because they have high concentrations of the bright yellow toxin vulpinic acid.

Wolf lichen (*Letharia vulpicida*) was used in Scandinavia to poison wolves. The process begins by adding the lichens to various baits such as reindeer blood and other meats, while sometimes mixing the concoction with ground glass or strychnine. Wolves that ate the concoction were reported to succumb in less than 24 hours. The Achomawi people of northern California use *Letharia* to poison arrowheads. The arrowheads would be soaked in the lichens for a year sometimes with the addition of rattlesnake venom. Although toxic, wolf lichens were used to treat sores and inflammation by indigenous people in north California and southern British Columbia, and even taken internally as medicine.

Lichens for Food

There are records of lichens being used as food by many different human cultures across the world. Lichens are eaten by people in North America, Europe, Asia, and Africa, and perhaps elsewhere. Often lichens are merely famine foods eaten in times of dire needs, but in some cultures lichens are a staple food or even a delicacy. Two problems often encountered with eating lichens is that they usually contain mildly toxic secondary compounds, and that lichen polysaccharides are generally indigestible to humans. Many human cultures have discovered preparation techniques to overcome these problems. Lichens are often thoroughly washed, boiled, or soaked in ash water to help remove secondary compounds.

In the past Iceland moss (*Cetraria islandica*) was an important human food in northern Europe and Scandinavia, and was cooked in many different ways, such as bread, porridge, pudding, soup, or salad. Wila (*Bryoria fremontii*) was an important food in parts of North America, where it was usually pitcooked. It is even featured in a Secwepemc story. Reindeer lichen (*Cladina* spp.) is a staple food of reindeer and caribou in the Arctic. Northern peoples in North America and Siberia traditionally eat the partially digested lichen after they remove it from the rumen of caribou that have been killed. It is often called 'stomach icecream'. Rock tripe (*Umbilicaria* spp. and *Lasalia* spp.) is a lichen that has frequently been used as an emergency food in North America. One species of *Umbilicaria*, Iwa-take (*U. esculenta*), is used in a variety of traditional Korean and Japanese foods. It is quite expensive, and is collected off the sides of cliffs. In India, and other centres of curry powder production, garam masala sauce contains certain lichens used as bulking agents.

Very few lichens are poisonous. Poisonous lichens include those high in vulpinic acid or usnic acid. Most (but not all) lichens that contain vulpinic acid are yellow, so any yellow lichen should be considered to be potentially poisonous.

Lichens for Embalming

Embalming began in Egypt about 5,000 years ago, and the use of lichens in embalming can be seen dating back to the 18th Dynasty where *Pseudevernia furfuracea* was found in an Egyptian vase. The process of using lichens in embalming is a fascinating discovery. For starters, to embalm a body, a slit was made in the abdomen and the organs and viscera were removed. The organs and viscera were wrapped in separate linen packets and replaced in the body or put in canopic jars between the legs. The body cavity was then packed with lichen, sawdust, bruised myrrh, cassia, and every other spices. *Pseudevernia furfuracea* was employed due to its preservative and aromatic qualities. Also it was used simply as a highly absorbent, light-weight packaging material. It also contains antibiotic substances. These qualities helped inhibit bacterial decay of the mummies. Another discovery is that the Egyptians would grind and mix *Pseudevernia furfuracea* with their flour for bread. The bread was then placed with the mummy and thought to be the first meal for the mummy in its afterlife. *Pseudevernia furfuracea* was imported shiploads from the Grecian archipelago to

Alexandria. Whether *Pseudevernia furfuracea*, was used for embalming, medicinal purposes, or as an ingredient in bread, it has its place in history as an important plant connecting a culture to its environment. Today, embalming fluids are coloured from the lichen dye orchil into a product called Cudbear, illustrating how a historical procedure can influence future practices.

Other Human Uses of Lichens

Lichens have been and are still being used for many other purposes, including

- Alcohol production (for fermentable carbohydrates, as catalysts, and/or as flavour/preservatives)
- Cosmetics (for hair, and/or sweet smelling powders)
- Perfumes
- Decorations (including costumes and artwork)
- Fibre (clothing, housing, cooking, sanitation)
- Animal feed (both fodder and forage)
- Fuel
- Industrial purposes (production of acid, antibiotic, carbohydrate, litmus)
- Tanning
- Hunting/fishing (to find prey, or to lure them in)
- Navigation
- Insect repellent/insecticide
- Preservatives (for food or beer)
- Poison (arrowheads, wolves)
- Mummies
- Rituals
- Magic.
- Tobacco
- Narcotics
- Hallucinogens

Rust (Fungus)

Rusts are plant diseases caused by pathogenic fungi of the order Pucciniales (previously also known as Uredinales). An estimated 168 rust genera and approximately 7000 species, more than half of which belong to the genus *Puccinia,* are currently accepted. Rust fungi are highly specialized parasites with several unique features. A single species may produce up to five morphologically and cytologically distinct spore-producing structures viz., spermagonia, aecia, uredinia, telia, and basidia in successive stages of reproduction. Unlike other plant pathogens, rust usually affects healthy and vigorously growing plants, so the infection is limited to plant parts, such as leaves, petioles, tender shoots, stem, fruits, etc. perennial systemic infection may cause deformities such as growth retardation, witches brooms, stem canker, hypertrophy of the affected tissues or formation of galls.

Plants with severe rust infection may appear stunted, chlorotic (yellowed), or otherwise discoloured. Rust sporulates on affected plant parts. Rust is most commonly seen as coloured powder, composed of tiny aeciospores which land on vegetation producing pustules, or uredia, that form on the lower surfaces. During late spring or early summer, yellow orange or brown, hairlike or ligulate structures called telia grow on the leaves or emerge from bark of woody hosts such as *Juniperus* species. These telia produce teliospores which will germinate into aerial basidiospores, spreading and causing further infection.

Taxonomy

The taxonomy of Pucciniales is complex and the darker coloured smuts can be mistaken for rust. Rusts are so named after the reddish rusty looking sori and the disease is usually noticed after rain.

The group is considered one of the most harmful pathogens to agriculture,horticulture and forestry. These fungi are major concerns and limiting factors for successful cultivation of agricultural and forestry crops. White pine blister rust, wheatstem rust, and coffee rust are examples of notoriously damaging, economically important rusts.

Spores

Rusts can produce up to five spores types during their life cycle

- 0-Pycniospores (Spermatia)-Haploid gametes in heterothallic rusts.
- I-Aeciospores-non-repeating dikaryotic vegetative spores
- II-Urediniospores-repeating dikaryotic vegetative spores. These spores are referred to as the repeating stage because they can cause auto-infection (re-infect the same host from which the spores were borne). These spores are red/orange and are a characteristic sign of rust fungus infection.

- III-Teliospores-Diploid spores that produce basidia and are the survival stage of life cycle
- IV-Basidiospores-stem from basidia. Haploid spores which infect the alternate host. although these are rarely observed outside of the laboratory.

Rust fungi can be categorized by how many types of spores are produced during the life cycle. Fungi that produce all five spores (sometimes excluding pycniospores) are termed macrocyclic. Fungi that lack pycniospores and aeciospores in their life cycle are termed microcyclic and always have an autoecious life cycle. Demicyclic fungi delete the uredial (repeating) stage from the life cycle. Understanding the life cycles of rust fungi allows for proper disease management.

Life Cycle

All rusts are obligate parasites, meaning that they require a living host to complete their life cycle. They generally do not kill the host plant but can severely reduce growth and yield. Cereal crops can be devastated in one season and trees that get infected in the main stem within the first five years, invariably die.

Rust fungi can also be categorized by their life cycle. Three basic types of life cycles are recognized based on theb number of spore states as macrocyclic, demicyclic, and microcyclic. The macrocyclic life cycle has all spore states, the demicyclic lacks the uredinial state, and the microcyclic cycle lacks both the aecial and uredinial states, thus possess only spermagonia and telia. Spermagonia may be absent from each type but especially the microcyclic life cycle. In macrocyclic and demicyclic life cycles, the rust may be either host alternating (heteroecious), i.e., the aecial state is on one kind of plant but the telial state on a different and unrelated plant, or non-host alternating (autoecious), i.e., the aecial and telial states on the same kind of plant. Heteroecious rust fungi require two unrelated hosts to complete their life cycle, with the primary host being infected by aeciospores and the alternate host being infected with basidiospores. This can be contrasted with an autoecious fungus which can complete its life cycle on a single host species.

Host Plant-rust Fungus Relationship

There are definite patterns of relationship with host plant groups and the rust fungi that parasitize them. Some genera of rust fungi, especially *Puccinia* and *Uromyces*, comprise species that are capable of parasitizing plants of many families. But many genera appear to be rather definitely restricted to certain plants. Host restriction may, in heteroecious species, apply to both phases of life cycle or to only one phase.

Infection Process

The fungi produce asexual spores which disperse by wind, water or by insect vectors spreading the infection.

Rust fungi are biotrophs taking nutrients from living cells. When airborne spores settle on a plant, weak hydrophobic interactions are formed with the cutin on the plant cell surface, securing it. By a process not fully understood, the production of mucous like substances called 'adhesins', initially stick the spore to the plant surface.

Once attached, the spore germinates by growing a germ tube and eventually locates a stoma by a touch responsive process known as thigmotropism. This involves growing towards a ridge between the epidermal cells, followed by a perpendicular growth which end at the stoma.

Inside the stoma, the hyphae tips flatten out to form structures known as appresorium that lock to the cell walls. It is thought that the whole process is mediated by a mechano-sensitive calcium ion channel, located within the germ tube tip, which produces electric currents that stretch the cell membranes, changing gene expression and forming the appresorium.

Then a peg grows into the plant's mesophyll cells. The peg produces specialised hyphal tips, known as haustorium. These spread around the plant cells without invading the membranes. The plant cell membranes invaginate around the main haustorial body forming a space known as the extra-haustorial matrix. An iron and phosphorus rich neck band bridges the plant and fungal membranes in the space between the cells for water flow, known as the apoplast, thus preventing the nutrients reaching the plant's cells. The haustorium contains amino acid and hexose sugar transporters and H^+-ATPases which are used for active transport of nutrients from the plant nourishing the fungus. It continues growing until spore growth occurs. The process repeats every 10 – 14 days, producing numerous spores, carried by wind to new hosts.

Common Rust Fungi in Agriculture

- *Puccinia sorghi* causes common rust in corn
- *Gymnosporangium juniperi-virginianae* (Cedar-apple rust); the *Juniperus virginiana* is the primary (telial) host and the apple,

pear or hawthorn is the secondary (aecial) host.Heteroecious and demicyclic

- *Cronartium ribicola* (White pine blister rust); the primary host are currants, and white pines the secondary. Heterocyclic and macrocyclic
- *Hemileia vastatrix* (Coffee rust); Primary host is coffee plant; Unknown alternate host. Heteroecious
- *Puccinia graminis* (Stem rust of wheat and Kentucky bluegrass); Primary hosts include: Kentucky bluegrass, barley, and wheat; Common barberry is the alternate host. Heteroecious and macrocyclic
- *Puccinia coronata* (Crown Rust of Oats and Ryegrass); Oats are the primary host; *Rhamnus spp.* (Buckthorn) is alternate host. Heteroecious and macrocyclic
- *Phakopsora meibomiae* and *P. pachyrhizi* (Soybean rust); Primary host is soybean and various legumes. Unknown alternate host. Heteroecious
- *Uromyces phaseoli* (Bean rust); Primary host:bean. Autoecious and macrocyclic
- *Puccinia hemerocallidis* (Daylily rust); Daylily is primary host; *Patrina sp* is alternate host. Heteroecious and macrocyclic
- *Puccinia persistens subsp. triticina* causes wheat rust in grains. It is also known as 'brown or red rust'.
- *Puccinia striiformis* causes 'Yellow rust' and *Puccinia graminis* causes 'black rust' in cereals.
- *Uromyces appendeculatus* is a problem on beans

Management of Rust Fungi Diseases

The control methods of rust fungus diseases depend largely on the life cycle of the particular pathogen. The following are examples of disease management plans used to control macrocyclic and demicyclic diseases:-

Macrocyclic Disease: Developing a management plan for this type of disease depends largely on whether the repeating stage (urediniospores) occur on the economically important host plant or the alternate host. For example, the repeating stage in white pine blister rust disease does not occur on white pines but on the alternate host, *Ribes spp.* During August and September *Ribes spp.* give rise to teliospores which infect white pines. Removal of the alternate host disrupts the life cycle of the rust fungi *Cronartium ribicola*, preventing the formation of basidiospores which infect the primary host. Although spores from white pines cannot infect other white pines, survival spores may overwinter on infected pines and reinfect *Ribes spp.* the following season. Infected tissue is removed from white pines and strict quarantines of *Ribes spp.* are maintained in high risk areas.

Puccinia graminis is a macrocyclic heteroecious fungus that causes wheat stem rust disease. The repeating stage in this fungus occurs on wheat and not the alternate host, barberry. The repeating stage allows the disease to persist in wheat even though the alternate host may be removed. Planting resistant crops is the ideal form of disease prevention, however, mutations can give rise to new strains of fungi that can overcome plant resistance. Although the disease cannot be stopped by removal of the alternate host, the life cycle is disrupted and the rate of mutation is decreased because of reduced genetic recombination. This allows resistance bred crops to remain effective for a longer period of time.

Demicyclic Disease: Because there is no repeating stage in the life cycle of demicyclic fungi, removal of the primary or the alternate host will disrupt the disease cycle. This method, however, is not highly effective in managing all demicyclic diseases. Cedar-apple rust disease, for example, can persist despite removal of one of the hosts since spores can be disseminated from long distances. The severity of Cedar-apple rust disease can be managed by removal of basidiospore producing galls from junipers or the application of protective fungicides to junipers.

Home Control

Rust is very hard to treat. Fungicides such as Mancozeb or Triforine may help but may never eradicate the disease. Some organic preventative solutions are available and sulphur powder is known to stop germination. High standards of hygiene and good soil drainage and careful watering may minimise problems. Any appearance of rust must be immediately dealt with by removing and burning all affected leaves. Composting, or leaving infected vegetation on the ground will spread the disease.

Commercial Control

In large plantations in USA, fungicides are applied by air. The process is expensive and fungicide application is best reserved for seasons when foliar diseases are severe. Research indicates, the higher the foliar disease severity, the greater the return from the use of fungicides. There are a variety of preventative methods that can also be employed.

- Symptoms of rust disease are correlated to relatively high moisture. The avoidance of overhead watering at night, using drip irrigation, reducing crop density, and using fans to circulate air flow will lower the relative moisture and decrease the severity of rust infection.
- The use of rust resistant plants
- Crop rotation can break the disease cycle because many rusts are host specific.
- Inspect all imported plants and cuttings for symptoms. It is important to continuously observe these plants because rust diseases have a latent period (plant has the disease but shows no symptoms).
- Many crops, such as wheat, are replanted with disease-free seed.

Host Plants Affected

Rusts are often named after the host species that they infect. For example; *Puccinia xanthii* infects the flowering plant Cocklebur (*Xanthium*).

It is probable that most plant species are affected by some species of rust.

Recently, a total of 95 rust fungi belonging to 25 genera associated with 117 forest plant species belonging to 80 host genera under 43 host families were reported from the Western Ghats, Kerala, India. Rust fungi include:

Aecidium, *Cerotelium*,*Chaconia*, *Coleosporium*, *Crossopsora*, *Didymopsorella*, *Hamspora*, *Hapalophragmidium*, *Hemileia*, *Kernkampella*, *Kuehneola*, *Kweilingia*, *Macabuna*, *Maravalia*, *Melampsora*, *Olivea*, *Physopella*, *Puccinia*, *Ravenelia*, *Uraecium*, *Úredo*, *Uredopeltis*, *Uromyces*, *Xenostele*, *Zaghouania*. Rust infected host genera include; *Acacia*, *Acalypha*, *Ageratina*, *Albizia*, *Arundinaria*, *Bambusa*, *Bidens*,*Blepharis*, *Bombax*, *Bridelia*, *Callicarpa*, *Canarium*, *Canthium*, *Catunaragam*, *Cinnamomum*, *Cissus*, *Cleistanthus*, *Clerodendron*, *Coffea*, *Coix*, *Cosmostigma*, *Crotalaria*, *Dalbergia*, *Dendrocalamus*, *Derris*, *Diospyros*, *Dipterocanthus*, *Elaeagnus*, *Elephantopus*, *Elettaria*, *Eragrostis*, *Euphorbia*, *Ficus*, *Flacourtia*,*Grewia*, *Holarrhena*, *Holoptelia*, *Hypericum*, *Ichnocarpus*,

*Ischaemum, Jasminum, Justicia, Loesneriella, Luvunga, Meiogyne, Meliosma, Mimusops, Morus, Neolitzea, Ocimum, Olea, Oxalis, Pavetta, Persicaria, Phyllanthus, Plectranthus, Plumeria, Pongamia, Premna, Protasparagus, Rubus, Salix, Spondia, Strobilanthes, Strychnos, Tabernaemontana, Terminalia, Toddalia, Trichosanthes, Vernonia, Vigna, Wrightia, Xanthophyllum, Xylia,*and *Ziziphus.* Some of the better known hosts include

- *Arisaema triphyllum*, Jack-in-the-pulpit
- *Avena sativa*, Oats
- *Berberis vulgaris*, Common barberry
- Broad beans
- *Coffea arabica* - Coffee
- *Crataegus monogyna* - Hawthorn
- Chrysanthemum
- *Cydonia* - Quince
- *Euphorbia maculata*, Spotted Spurge
- *Fuchsia spp*, Fuchsia
- Garlic
- *Hordeum vulgare*, Barley
- *Juniperus virginianae*, Red Cedar (Juniper apple disease)
- *Juniperus communis* - Juniper
- *Allium ampeloprasum* - Leek
- *Malus* – Apple
- *Mespilus* - Medlar
- Onion
- Pelargonium
- *Primula veris*
- *Primula vulgaris*
- *Pyrus* - Pear
- *Rosa spp*, Roses
- *Triticum spp.*, Wheat
- *Oxalis spp.*, Oxalis
- *Secale cereale*, Rye
- *Senecio vulgaris* -Common groundsel
- *Xanthium canadense*Cocklebur

Erysiphales

Erysiphales are an order of ascomycete fungi. The order contains one family, Erysiphaceae. Many of them cause plant diseases called powdery mildew.

Systematics

The order contains one family (Erysiphaceae), 28 genera and approximately 100 species. Many imperfect fungi (fungi whose sexual reproduction is unknown) belong here, especially the genus *Oidium*. Recent molecular data have revealed the existence of six main evolutionary lineages. Clade 1 consists of *Erysiphe*, *Microsphaera*, and *Uncinula*, all of which have an *Oidium* subgenus *Pseudoidium* mitosporic state. Clade 2 consists of *Erysiphe galeopsidis* and *Erysiphe cumminsiana* (anamorphs in *Oidium* subgenus *Striatoidium*). Clade 3 consists of *Erysiphe* species with anamorphs in *Oidium* subgenus *Reticuloidium*. Clade 4 consists of *Leveillula* and *Phyllactinia*, which have *Oidiopsis* and *Ovulariopsis* mitosporic states, respectively. Clade 5 consists of *Sphaerotheca*, *Podosphaera*, and *Cystotheca*, which have *Oidium* subgenera *Fibroidium* and *Setoidium* mitosporic states. Clade 6 consists of *Blumeria graminis*, which has an *Oidium* subgenus *Oidium* mitosporic state. Several morphological characters have been analysed and found not to conflict with the molecular data.

Characteristics

- Erysiphales have superficial mycelium which extracts nourishment from the host plant through specialized hyphae that penetrate the epidermal cells of the host by means of absorbing organs called haustoria.
- The teleomorphs are usually more distinctive and diverse than the anamorphs. Whether the asci are bitunicate or unitunicate (i.e. consisting of one or two layers), is as yet a matter of discussion.
- The cleistothecia have the asci arranged in a hymenial layer, resembling perithecia.

The cleistothecia (chasmothecia) are minute, usually not much more than 0.1 mm in diameter. From the outer wall of the cleistothecium specialised hyphae (appendages) grow out. The number of asci per ascoma varies, and is important in discriminating between genera.

- *Erysiphales* are notable for intricateappendages which follow fractal geometry within Fibonacci numbers and can be useful for species identification.

Life Cycle

The infection of the host plant begins with the sexual ascospores, or the asexual conidia germinating on the surface of the plants leaf or stem, resulting in septate mycelium of uninucleate cells. In most powdery mildews only the epidermal cells are attacked. The external mycelium gives rise to short, erect conidiophores, each of which bearing a single row of barrel-shaped spores, the youngest being at the base (the affected parts become thus covered with a forest of conidiophores assuming a white powdery appearance).

The ripe spores become detached and are readily dispersed by the wind, causing fresh infection. In autumn the sexual cleistothecia are produced. The cleistothecia represent the resting (hibernating) stage of the pathogen. The ascospores remain dormant all winter to germinate in spring. When the asci expand they rupture the chasmothecia wall, throwing the ascospores into the air.

Ecology

Erysiphales are obligate parasites on leaves and fruits of higher plants, causing diseases called powdery mildews. Most attempts to grow them in culture have failed.

Lycoperdon

Lycoperdon is a genus of puffball mushrooms. The genus has a widespread distribution and contains about 50 species. In general, it contains the smaller species such as the pear shaped puffball and the gem studded puffball. It was formerly classified within the now-obsolete order Lycoperdales, as the type genus which, following a restructuring of fungal taxonomy brought about by molecular phylogeny, has been split. *Lycoperdon* is now placed in the family Agaricaceae of the order Agaricales.

The name comes from *lycos* meaning *wolf* and *perdon* meaning *to break wind*; thus the name literally means *wolf-farts*.

Species

- *Lycoperdon bovista* Sowerby 1803
- *Lycoperdon caudatum* (Syn. *Lycoperdon pedicellatum*) J.Schröt.
- *Lycoperdon curtisii*
- *Lycoperdon echinatum* Pers.
- *Lycoperdon ericaceum* (Syn. *Lycoperdon muscorum*)
- *Lycoperdon ericaceum*, var. *subareolatum*

- *Lycoperdon lambinonii* Demoulin 1972
- *Lycoperdon lividum* (Syn. *Lycoperdon spadiceum*) Pers. 1809
- *Lycoperdon mammiforme* Pers. 1801
- *Lycoperdon marginatum* (Syn. *Lycoperdon candidum*) Vittadini
- *Lycoperdon molle*
- *Lycoperdon nigrescens* (Syn. *Lycoperdon foetidum*) Bonord
- *Lycoperdon norvegicum* Demoulin 1971
- *Lycoperdon ovoidisporum*
- *Lycoperdon perlatum* (Syn. *Lycoperdon gemmatum*)
- *Lycoperdon pyriforme* Schaeff. ex Pers. (now *Morganella pyriformis*)
- *Lycoperdon pulcherrimum* Berk. & M.A. Curtis
- *Lycoperdon rupicola*
- *Lyoperdon subumbrinum*
- *Lycoperdon umbrinum*

Saprolegnia

Saprolegnia is a genus of freshwater mould often called a "cotton mould" because of the characteristic white or grey fibrous patches it forms. Current taxonomy puts Saprolegnia as a genus of the heterokonts in the order Saprolegniales.

Habits

Saprolegnia, like most water moulds, is both a saprotroph and necrotroph. Typically feeding on waste from fish or other dead cells, they will also take advantage of creatures that have been injured or compromised eggs. When they inhabit a live animal, they exhibit as a fungal infection known as mycoses.

Saprolegnia is tolerant to a wide range of temperature, 3°C to 33°C, but is more prevalent in lower temperatures. While it is found most frequently in freshwater, it will also tolerate brackish water and even moist soil.

Saprolegnia filaments (hyphae) are long with rounded ends, containing the zoospores. Saprolegnia generally travels in colonies consisting of one or more species. They first form a mass of individual hyphae. When the mass of hyphae grows large enough in size to be seen without use of a microscope, it can be called a mycelium. Colonies are generally white in colour, though they may turn grey under the precesence of bacteria or other debris which has become caught in the fibrous mass.

Reproduction

It has a diploid life cycle which includes both sexual and asexual reproduction. In the asexual phase, a spore of *Saprolegnia* releases zoospores. Within a few minutes, this zoospore will encyst, germinate and release another zoospore. This second zoospore has a longer cycle during which most dispersal happens; it will continue to encyst and release a new spore in a process called polyplanetism until it finds a suitable substrate.

When a suitable medium is located, the hairs surrounding the spore will lock onto the substrate so that the sexual reproduction phase can start. It is also during this stage of polyplanetism that the *Saprolegnia* are capable of causing infection; the most pathogenic species have tiny hooks at the end of their hairs to enhance their infectious ability.

Once firmly attached, sexual reproduction begins with the production of male and female gametangium, antheridia and oogonium respectively. These unite and fuse together via fertilization tubes. The zygote produced is named an oospore.

Characteristics of Infection

Saprolegnia is generally a secondary pathogen, though in the right circumstances, it can act as primary. It most frequently targets fish, both in the wild and in tank environments. Through cellular necrosis and other epidermal damage, *Saprolegnia* will spread across the surface of its host as a cotton-like film.

Though it often stays in the epidermal layers, the mould does not appear to be tissue specific. A *Saprolegnia* infection is usually fatal, eventually causing haemodilution, though the time to death varies depending on the initial site of the infection, rate of growth and the ability of the organism to withstand the stress of the infection.

The extensive mortalities of salmon and migratory trout in the rivers of western Europe in the 1970s and 1980s in the UDN outbreak were probably almost all ultimately caused by the secondary *Saprolegnia* infections.

Historial evidence suggest the *Saprolegnia* species affecting Australian freshwater fish may be an introduced strain, imported in the 1800s with exotic fish species.

Agaricales

The fungal order Agaricales, also known as gilled mushrooms (for their distinctive gills) or euagarics, contains some of the most familiar types of mushrooms. The order has 33 extant families, 413 genera, and over 13000 described species, along with five extinct genera known only from the fossil record. They range from the ubiquitous common mushroom to the deadly destroying angel and the hallucinogenic fly agaric to the bioluminescent jack-o-lantern mushroom.

History, Classification and Phylogeny

In his three volumes of *Systema Mycologicum* published between 1821 and 1832, Elias Fries put almost all of the fleshy, gill-forming mushrooms in the genus *Agaricus*. He organized the large genus into "tribes", the names of many of which still exist as common genera of today. Fries later elevated several of these tribes to generic level, but later authors—including Gillet, Karsten, Kummer, Quélet, and Staude—made most of the changes. Fries based his classification on macroscopic characters of the fruit bodies and colour of the spore print. His system had been widely used as it had the advantage that many genera could be readily identified based on characters observable in the field. Fries's classification was later challenged when microscopic studies of basidiocarp structure, initiated by Fayod and Patouillard, demonstrated several of Fries's groupings were unnatural. In more recent history, Rolf Singer's influential work *The Agaricales in Modern Taxonomy*, published in four editions spanning from 1951 to 1986, used both Fries's macroscopic characters and Fayod's microscopic characters to reorganize families and genera; his most recent classification included 230 genera within 18 families. Singer treated three major groups within the Agaricales *sensu lato*: the Agaricales *sensu stricto*, Boletineae, and Russulales. These groups are still accepted by modern treatments based on DNA analysis, as the euagarics clade, bolete clade, and russuloid clade.

Molecular phylogenetics research has demonstrated that the euagarics clade is roughly equivalent to Singer's Agaricales *sensu stricto*. A recent (2006) large-scale study by Brandon Matheny and colleagues used nucleic acid sequences representing six gene regions from 238 species in 146 genera to explore the phylogenetic grouping within the Agaricales. The analysis showed that most of the species tested could be grouped into six clades that were named the Agaricoid, Tricholomatoid, Marasmioid, Pluteoid, Hygrophoroid and Plicaturopsidoid clades.

Some notable fungi with gill-like structures, such as chanterelles, have long been recognized as being substantially different from usual Agaricales. Interestingly, molecular studies are showing more groups of agarics as being more divergent than previously thought, such as the genera *Russula* and *Lactarius* belonging to a separate order Russulales, and other gilled fungi, including such species as *Paxillus involutus* and *Hygrophoropsis aurantiaca* showing a closer affinity with Boletes in the order Boletales.

Also, some other quite distinctive fungi, the puffballs, and some clavaroid fungi, e.g. *Typhula*, and the Beefsteak fungus have been recently been shown to lie within the Agaricales.

The term agaric had traditionally referred to Agaricales, which were defined as exactly those fungi with gills. Given the discoveries described above, those two categories are not synonymous (although there is a very large overlap between the two groups).

Distribution and Habitat

Figure: Cortinarius archeri

Agarics are ubiquitous, being found across all continents. Most are terrestrial, their habitats including all types of woodland and grassland, varying largely from one genus to another. Agarics were long thought to be solely terrestrial, until the 2005 discovery of *Psathyrella aquatica*, the only gilled mushroom known to fruit underwater.

Agaricals are known from five monotypic genera found fossilized in amber. The oldest records are from two Cretaceous age genera; the Albian age (approximately 100 Ma) *Palaeoagaracites antiquus* from Burmese amber and the slightly younger Turonian New Jersey Amber species *Archaeomarasmius leggeti*. The three other species, *Aureofungus yaniguaensis*, *Coprinites dominicana* and *Protomycena electra* are known from single specimens found in the Dominican amber mines of Hispaniola.

Characteristics

Basidiocarps of the agarics are typically fleshy, with a stipe, often called a stem or stalk, a pileus (or cap) and lamellae (or gills), where basidiospores are produced. This is the stereotypical structure of a mushroom.

Life Cycle

The fungus fruit body is the spore-producing stage of the life cycle. Most fungi reproduce by spores and the fruit bodies are developed specifically for the production and dispersal of spores. The spores produced by fruit bodies are usually the result of sexual reproduction.

The fruit body is the visible part of the growing fungus. It is supported by and develops from an extensive network of thread-like filaments called hyphae. Hyphae are often collectively termed the mycelium; the food-absorbing part of the fungus—as opposed to the spore-producing fruit body of the fungus—is called the vegetative mycelium. The individual hyphae that compose the mycelium absorb nutrients and water from the substratum in which they are growing. When the nutrient supply is adequte and environmental conditions are favourable, some fungi may grow in the same location for several years. Fungi cannot make their own food, namely carbohydrates, as can green plants. Some species are saprobic, obtaining nutrients from dead organic material, whereas others are parasitic on living plants or animals or even on other fungi. Many fungi, especially gilled mushroomes and boletes, have an extensive mycelium that lives in association with the roots of woody plants. This association, which is beneficial to both the fungus and host plant, is termed a mycorrhiza.

When the environmental conditions are favourable and the mycelium is at the proper stage of development, one or more fruit bodies are produced by the fungus. The actual conditions necessary for fruit body formation and spore production are not clearly understood. Humidity, light, temperature, aeration, and nutrition are all factors thought to be important in fruit body formation. The genetic makeup and the general physiology of the fungus hyphae are also important in the initiation and formation of young fruit bodies and their development to a mature stage. The spores produced by a fruit body are released when it is mature. When they land in a suitable environment, the spores germinate and the hyphae grow to initiate the life cycle anew.

Genera Incertae Sedis

Figure: Setchelliogaster *is a genus of gasteroid fungi in the Agaricales that is* incertae sedis *with respect to familial placement.*

There are several genera classified in the Agaricales that are i) poorly known, ii) have not been subjected to DNA analysis, or iii) if analysed phylogenetically do not group with as yet named or identified families, and have not been assigned to a specific family (i.e., *Incertae sedis* with respect to familial placement). These include:

- *Atheniella* Redhead, Moncalvo, Vilgalys, Desjardin & B.A.Perry (2012)
- †*Aureofungus* Hibbett, Manfr. Binder & Zheng Wang (2003)
- *Brunneocorticium* Sheng H.Wu (2007)
- *Cheilophlebium* Opiz & Gintl (1856)
- *Cleistocybe* Ammirati, A.D.Parker & Matheny (2007)
- *Cribrospora* Pacioni & P.Fantini (2000)
- *Disporotrichum* (anamorph) Stalpers (1984)
- *Mesophelliopsis* Bat. & A.F.Vital (1957)
- †*Palaeoagaracites* Poinar & Buckley (2007)
- *Panaeolina* Maire (1933)
- *Panaeolus* (Fr.) Quél. (1872)
- *Phlebophyllum* R.Heim (1969)
- *Plicatura* Peck (1872)
- *Plicaturopsis* D.A.Reid (1964)
- *Sedecula* Zeller (1941)
- *Setchelliogaster* Pouzar (1958)
- *Trichocybe* Vizzini (2010)

Fossil

Fossils (from Classical Latin *fossilis,* literally "obtained by digging") are the preserved remains or traces of animals (also known as *zoolites*), plants, and other organisms from the remote past. The totality of fossils, both discovered and undiscovered, and their placement in *fossiliferous* (fossil-containing) rock formations and sedimentary layers (strata) is known as the *fossil record.*

The study of fossils across geological time, how they were formed, and the evolutionary relationships between taxa (phylogeny) are some of the most important functions of the science of paleontology. Such a preserved specimen is called a "fossil" if it is older than some minimum age, most often the arbitrary date of 10,000 years. Hence, fossils range in age from the youngest at the start of the Holocene Epoch to the oldest from the Archaean Eon, up to 3.4 billion years old. The observation that certain fossils were associated with certain rock strata led early geologists to recognize a geological timescale in the 19th century. The development of radiometric dating techniques in the early 20th century allowed geologists to determine the numerical or *"absolute" age* of the various strata and thereby the included fossils.

Like extant organisms, fossils vary in size from microscopic, even single bacterial cells one micrometer in diameter, to gigantic, such as dinosaurs and trees many metres long and weighing many tons. A fossil normally preserves only a portion of the deceased organism, usually that portion that was partially mineralized during life, such as the bones and teeth of vertebrates, or the chitinous or calcareous exoskeletons of invertebrates. Fossils may also consist of the marks left behind by the organism while it was alive, such as the footprint or feces (coprolites) of a reptile. These types of fossil are called trace fossils (or *ichnofossils*), as opposed to *body fossils*. Finally, past life leaves some markers that cannot be seen but can be detected in the form of biochemical signals; these are known as *chemofossils* or biomarkers.

Fossilization Processes

Fossilization processes proceed differently according to tissue type and external conditions.

Permineralization

Permineralization is a process of fossilization that occurs when an organism is buried. The empty spaces within an organism (spaces filled with liquid or gas during life) become filled with mineral-rich groundwater. Minerals precipitate from the groundwater, occupying the empty spaces. This process can occur in very small spaces, such as within the cell wall of a plant cell. Small scale permineralization can produce very detailed fossils. For permineralization to occur, the organism must become covered by sediment soon after death or soon after the initial decay process. The degree to which the remains are decayed when covered determines the later details of the fossil. Some fossils consist only of skeletal remains or teeth; other fossils contain traces of skin, feathers or even soft tissues. This is a form of diagenesis.

Casts and Molds

In some cases the original remains of the organism completely dissolve or are otherwise destroyed. The remaining organism-shaped hole in the rock is called an *external mold*. If this hole is later filled with other minerals, it is a *cast*. An endocast or *internal mold* is formed when sediments or minerals fill the internal cavity of an organism, such as the inside of a bivalve or snail or the hollow of a skull.

Authigenic Mineralization

This is a special form of cast and mold formation. If the chemistry is right, the organism (or fragment of organism) can act as a nucleus for the precipitation of minerals such as siderite, resulting in a nodule forming around it. If this happens rapidly before significant decay to the organic tissue, very fine three-dimensional morphological detail can be preserved. Nodules from the Carboniferous Mazon Creek fossil beds of Illinois, USA, are among the best documented examples of such mineralisation.

Replacement and Recrystallization

Replacement occurs when the shell, bone or other tissue is replaced with another mineral. In some cases mineral replacement of the original shell occurs so gradually and at such fine scales that microstructural features are preserved despite the total loss of original material. A shell is said to be *recrystallized* when the original skeletal compounds are still present but in a different crystal form, as from aragonite to calcite.

Adpression (Compression-impression)

Compression fossils, such as those of fossil ferns, are the result of chemical reduction of the complex organic molecules composing the organism's tissues. In this case the fossil consists of original material, albeit in a geochemically altered state. This chemical change is an expression of diagenesis.

Often what remains is a carbonaceous film known as a phytoleim, in which case the fossil is known as a compression. Often, however, the phytoleim is lost and all that remains is an impression of the organism in the rock—an impression fossil. In many cases, however, compressions and impressions occur together. For instance, when the rock is broken open, the phytoleim will often be attached to one part (compression), whereas the counterpart will just be an impression. For this reason, one term coverrs the two modes of preservation: *adpression*.

Carbon Films

Carbon films are thin film coatings which consist predominantly of the chemical element carbon. The soft tissues of organisms are made largely of organic carbon compounds, leaving a thin film of carbon residue is left, forming a silhouette of the original organism called a carbon film.

Bioimmuration

Bioimmuration occurs when a skeletal organism overgrows or otherwise subsumes another organism, preserving the latter, or an impression of it, within the skeleton. Usually it is a sessile skeletal organism, such as a bryozoan or an oyster, which grows along a substrate, covering other sessile sclerobionts. Sometimes the bioimmured organism is soft-bodied and is then preserved in negative relief as a kind of external mold. There are also cases where an organism settles on top of a living skeletal organism and grows upwards, preserving the settler in its skeleton. Bioimmuration is known in the fossil record from the Ordovician to the Recent.

Fossil Record

Estimating Dates: Paleontology seeks to map out how life evolved across geologic time. A substantial hurdle is the difficulty of working out fossil ages. Beds that preserve fossils typically lack the radioactive elements needed for radiometric dating. This technique is our only means of giving rocks greater than about 50 million years old an absolute age, and can be accurate to within 0.5% or better. Although radiometric dating requires careful laboratory work, its basic principle is simple: the rates at which various radioactive elements decay are known, and so the ratio of the radioactive element to its decay products shows how long ago the radioactive element was incorporated into the rock. Radioactive elements are common only in rocks with a volcanic origin, and so the only fossil-bearing rocks that can be dated radiometrically are volcanic ash layers.

Stratigraphy

Consequently, palaeontologists rely on stratigraphy to date fossils. Stratigraphy is the science of deciphering the "layer-cake" that is the sedimentary record. Rocks normally form relatively horizontal layers, with each layer younger than the one underneath it. If a fossil is found between two layers whose ages are known, the fossil's age is claimed to lie between the two known ages. Because rock sequences are not continuous, but may be broken up by faults or periods of erosion, it is very difficult to match up rock beds that are not directly adjacent. However, fossils of species that survived for a relatively short time can be used to match isolated rocks: this technique is called *biostratigraphy*. For instance, the conodont *Eoplacognathus pseudoplanus* has a short range in the Middle Ordovician period. If rocks of unknown age have traces of *E. pseudoplanus*, they have a mid-Ordovician age. Such index fossils must be distinctive, be globally distributed and occupy a short time range to be useful. Misleading results are produced if the index fossils are incorrectly dated. Stratigraphy and biostratigraphy can in general provide only relative dating (*A* was before *B*), which is often sufficient for studying evolution. However, this is difficult for some time periods, because of the problems involved in matching rocks of the same age across continents. Family-tree relationships also help to narrow down the date when lineages first appeared. For instance, if fossils of B or C date to X million years ago and the calculated "family tree" says A was an ancestor of B and C, then A must have evolved earlier.

It is also possible to estimate how long ago two living clades diverged – i.e. approximately how long ago their last common ancestor must have lived – by assuming that DNA mutations accumulate at a constant rate. These "molecular clocks", however, are fallible, and provide only approximate timing: for example, they are not sufficiently precise and reliable for estimating when the groups that feature in the Cambrian explosion first evolved, and estimates produced by different techniques may vary by a factor of two.

Limitations

Organisms are only rarely preserved as fossils in the best of circumstances, and only a fraction of such fossils have been discovered. This is illustrated by the fact that the number of species known through the fossil record is less than 5% of the number of known living species, suggesting that the number of species known through fossils must be far less than 1% of all the species that have ever lived. Because of the specialized and rare circumstances required for a biological structure to fossilize, only a small percentage of life-forms can be expected to be represented in discoveries, and each discovery represents only a snapshot of the process of evolution. The transition itself can only be illustrated and corroborated by transitional fossils, which will never demonstrate an exact half-way point.

The fossil record is heavily slanted toward organisms with hard parts, leaving most groups of soft-bodied organisms with little to no role. It is replete with the vertebrates, the echinoderms, the brachiopods and some groups of arthropods.

Lagerstatten

Fossil sites with exceptional preservation—sometimes including preserved soft tissues—are known as Lagerstätten. These formations may have resulted from carcass burial in an anoxic environment with minimal bacteria, thus slowing decomposition. Lagerstätten span geological time from the Cambrian period to the present. Worldwide, some of the best examples of near-perfect fossilization are the Cambrian Maotianshan shales and Burgess Shale, the Devonian Hunsrück Slates, the Jurassic Solnhofen limestone, and the Carboniferous Mazon Creek localities.

Stromatolites

Stromatolites are layered accretionary structures formed in shallow water by the trapping, binding and cementation of sedimentary grains by biofilms of microorganisms, especially cyanobacteria. Stromatolites provide some of the most ancient fossil records of life on Earth, dating back more than 3.5 billion years ago.

Stromatolites were much more abundant in Precambrian times. While older, Archean fossil remains are presumed to be colonies of cyanobacteria, younger (that is, Proterozoic) fossils may be primordial forms of the eukaryote chlorophytes (that is, green algae). One genus of stromatolite very common in the geologic record is *Collenia*. The earliest stromatolite of confirmed microbial origin dates to 2.724 billion years ago.

A 2009 discovery provides strong evidence of microbial stromatolites extending as far back as 3.45 billion years ago.

Stromatolites are a major constituent of the fossil record for life's first 3.5 billion years, peaking about 1.25 billion years ago. They subsequently declined in abundance and diversity, which by the start of the Cambrian had fallen to 20% of their peak. The most widely supported explanation is that stromatolite builders fell victims to grazing creatures (the Cambrian substrate revolution), implying that sufficiently complex organisms were common over 1 billion years ago.

The connection between grazer and stromatolite abundance is well documented in the younger Ordovician evolutionary radiation; stromatolite abundance also increased after the end-Ordovician and end-Permian extinctions decimated marine animals, falling back to earlier levels as marine animals recovered. Fluctuations in metazoan population and diversity may not have been the only factor in the reduction in stromatolite abundance. Factors such as the chemistry of the environment may have been responsible for changes.

While prokaryotic cyanobacteria themselves reproduce asexually through cell division, they were instrumental in priming the environment for the evolutionary development of more complex eukaryotic organisms. Cyanobacteria (as well as extremophile Gammaproteobacteria) are thought to be largely responsible for increasing the amount of oxygen in the primeval earth's atmosphere through their continuing photosynthesis. Cyanobacteria use water, carbon dioxide and sunlight to create their food. A layer of mucus often forms over mats of cyanobacterial cells. In modern microbial mats, debris from the surrounding habitat can become trapped within the mucus, which can be cemented by the calcium carbonate to grow thin laminations of limestone. These laminations can accrete over time, resulting in the banded pattern common to stromatolites. The domal morphology of biological stromatolites is the result of the vertical growth necessary for the continued infiltration of sunlight to the organisms for photosynthesis. Layered spherical growth structures termed oncolites are similar to stromatolites and are also known from the fossil record. Thrombolites are poorly laminated or non-laminated clotted structures formed by cyanobacteria common in the fossil record and in modern sediments.

The Zebra River Canyon area of the Kubis platform in the deeply dissected Zaris Mountains of south western Namibia provides an extremely well exposed example of the thrombolite-stromatolite-metazoan reefs that developed during the Proterozoic period, the stromatolites here being better developed in updip locations under conditions of higher current velocities and greater sediment influx.

Types

Index: Index fossils (also known as guide fossils, indicator fossils or zone fossils) are fossils used to define and identify geologic periods (or faunal stages). They work on the premise that, although different sediments may look different depending on the conditions under which they were deposited, they may include the remains of the same species of fossil. The shorter the species' time range, the more precisely different sediments can be correlated, and so rapidly evolving species' fossils are particularly valuable. The best index fossils are common, easy-to-identify at species level and have a broad distribution—otherwise the likelihood of finding and recognizing one in the two sediments is poor

Trace

Trace fossils consist mainly of tracks and burrows, but also include coprolites (fossil feces) and marks left by feeding. Trace fossils are particularly significant because they represent a data source that is not limited to animals with easily fossilized hard parts, and they reflect animal behaviours. Many traces date from significantly earlier than the body fossils of animals that are thought to have been capable of making them. Whilst exact assignment of trace fossils to their makers is generally impossible, traces may for example provide the earliest physical evidence of the appearance of moderately complex animals (comparable to earthworms).

Coprolites are classified as trace fossils as opposed to body fossils, as they give evidence for the animal's behaviour (in this case, diet) rather than morphology. They were first described by William Buckland in 1829. Prior to this they were known as "fossil fir cones" and "bezoar stones." They serve a valuable purpose in paleontology because they provide direct evidence of the predation and diet of extinct organisms. Coprolites may range in size from a few millimetres to over 60 centimetres.

Transitional

A *transitional fossil* is any fossilized remains of a life form that exhibits traits common to both an ancestral group and its derived descendant group. This is especially important where the descendant group is sharply differentiated by gross anatomy and mode of living from the ancestral group. Because of the incompleteness of the fossil record, there is usually no way to know exactly how close a transitional fossil is to the point of divergence. These fossils serve as a reminder that taxonomic divisions are human constructs that have been imposed in hindsight on a continuum of variation.

Microfossils

Microfossil is a descriptive term applied to fossilized plants and animals whose size is just at or below the level at which the fossil can be analyzed by the naked eye. A commonly applied cutoff point between "micro" and "macro" fossils is 1 mm. Microfossils may either be complete (or near-complete) organisms in themselves (such as the marine plankters foraminifera and coccolithophores) or component parts (such as small teeth or spores) of larger animals or plants. Microfossils are of critical importance as a reservoir of paleoclimate information, and are also commonly used by biostratigraphers to assist in the correlation of rock units.

Resin

Fossil resin (colloquially called amber) is a natural polymer found in many types of strata throughout the world, even the Arctic. The oldest fossil resin dates to the Triassic, though most dates to the

Tertiary. The excretion of the resin by certain plants is thought to be an evolutionary adaptation for protection from insects and to seal wounds. Fossil resin often contains other fossils called inclusions that were captured by the sticky resin. These include bacteria, fungi, other plants, and animals. Animal inclusions are usually small invertebrates, predominantly arthropods such as insects and spiders, and only extremely rarely a vertebrate such as a small lizard. Preservation of inclusions can be exquisite, including small fragments of DNA.

Derived

A *derived, reworked* or *remanié fossil* is a fossil found in rock made significantly later than when the fossilized animal or plant died: it happens when a hard fossil is freed from a soft rock formation by erosion and redeposited in a currently forming sedimentary deposit.

Wood

Fossil wood is wood that is preserved in the fossil record. Wood is usually the part of a plant that is best preserved (and most easily found). Fossil wood may or may not be petrified. The fossil wood may be the only part of the plant that has been preserved: therefore such wood may get a special kind of botanical name. This will usually include "xylon" and a term indicating its presumed affinity, such as *Araucarioxylon* (wood of *Araucaria* or some related genus), *Palmoxylon* (wood of an indeterminate palm), or *Castanoxylon* (wood of an indeterminate chinkapin).

Subfossil

Subfossils refer to remains, such as bones, nests, or defecations, whose fossilization process is not complete, either for lack of time or because the conditions in which they were buried were not optimal for fossilization. Subfossils are often found in caves or other shelters where they can be preserved for thousands of years. The main importance of subfossil vs. fossil remains is that the former contain organic material, which can be used for radiocarbon dating or extraction and sequencing of DNA, protein, or other biomolecules. Additionally, isotope ratios can provide much information about the ecological conditions under which extinct animals lived. Subfossils are useful for studying the evolutionary history of an environment and can be important to studies in paleoclimatology.

Subfossils are often found in depositionary environments, such as lake sediments, oceanic sediments, and soils. Once deposited, physical and chemical weathering can alter the state of preservation.

Chemical Fossils

Chemical fossils are chemicals found in rocks and fossil fuels (petroleum, coal, and natural gas) that provide an organic signature for ancient life. Molecular fossils and isotope ratios represent two types of chemical fossils.

Pseudofossils

Pseudofossils are visual patterns in rocks that are produced by geologic processes rather than biologic processes. They can easily be mistaken for real fossils. Some pseudofossils, such as dendrites, are formed by naturally occurring fissures in the rock that get filled up by percolating minerals.

Other types of pseudofossils are kidney ore (round shapes in iron ore) and moss agates, which look like moss or plant leaves. Concretions, spherical or ovoid-shaped nodules found in some sedimentary strata, were once thought to be dinosaur eggs, and are often mistaken for fossils as well.

History of the Study of Fossils

Gathering fossils dates at least to the beginning of recorded history. The fossils themselves are referred to as the fossil record. The fossil record was one of the early sources of data underlying the study of evolution and continues to be relevant to the history of life on Earth. Paleontologists examine the fossil record to understand the process of evolution and the way particular species have evolved.

Explanations

Before Darwin: Many early explanations relied on folktales or mythologies. In China the fossil bones of ancient mammals including *Homo erectus* were often mistaken for "dragon bones" and used as medicine and aphrodisiacs. In the West fossilized sea creatures on mountainsides were seen as proof of the biblical deluge.

In 1027, the Persian Avicenna explained fossils' stoniness in *The Book of Healing*:

If what is said concerning the petrifaction of animals and plants is true, the cause of this (phenomenon) is a powerful mineralizing and petrifying virtue which arises in certain stony spots, or emanates suddenly from the earth during earthquake and subsidences, and petrifies whatever comes into contact with it. As a matter of fact, the petrifaction of the bodies of plants and animals is not more extraordinary than the transformation of waters.

Greek scholar Aristotle realized that fossil seashells from rocks were similar to those found on the beach, indicating the fossils were once living animals. Aristotle previously explained it in terms of vaporous exhalations, which Avicenna modified into the theory of petrifying fluids (*succus lapidificatus*), later elaborated by Albert of Saxony in the 14th century and accepted in some form by most naturalists by the 16th century.

More scientific views of fossils emerged during the Renaissance. Leonardo da Vinci concurred with Aristotle's view that fossils were the remains of ancient life. For example, da Vinci noticed discrepancies with the biblical flood narrative as an explanation for fossil origins:

> *"If the Deluge had carried the shells for distances of three and four hundred miles from the sea it would have carried them mixed with various other natural objects all heaped up together; but even at such distances from the sea we see the oysters all together and also the shellfish and the cuttlefish and all the other shells which congregate together, found all together dead; and the solitary shells are found apart from one another as we see them every day on the sea-shores.*
>
> *And we find oysters together in very large families, among which some may be seen with their shells still joined together, indicating that they were left there by the sea and that they were still living when the strait of Gibraltar was cut through. In the mountains of Parma and Piacenza multitudes of shells and corals with holes may be seen still sticking to the rocks...."*

William Smith (1769–1839), an English canal engineer, observed that rocks of different ages (based on the law of superposition) preserved different assemblages of fossils, and that these assemblages succeeded one another in a regular and determinable order. He observed that rocks from distant locations could be correlated based on the fossils they contained. He termed this the principle of *faunal succession.* This principle became one of Darwin's chief pieces of evidence that biological evolution was real.

Georges Cuvier came to believe that most if not all the animal fossils he examined were remains of extinct species. This led Cuvier to become an active proponent of the geological school of thought called catastrophism. Near the end of his 1796 paper on living and fossil elephants he said:

> *All of these facts, consistent among themselves, and not opposed by any report, seem to me to prove the existence of a world previous to ours, destroyed by some kind of catastrophe.*

Linnaeus and Darwin

Early naturalists well understood the similarities and differences of living species leading Linnaeus to develop a hierarchical classification system still in use today. Darwin and his contemporaries first linked the hierarchical structure of the tree of life with the then very sparse fossil record. Darwin eloquently described a process of descent with modification, or evolution, whereby organisms either adapt to natural and changing environmental pressures, or they perish.

When Darwin wrote *On the Origin of Species by Means of Natural Selection, or the Preservation of Favoured Races in the Struggle for Life*, the oldest animal fossils were those from the Cambrian Period, now known to be about 540 million years old. He worried about the absence of older fossils because of the implications on the validity of his theories, but he expressed hope that such fossils would be found, noting that: "only a small portion of the world is known with accuracy." Darwin also pondered the sudden appearance of many groups (i.e. phyla) in the oldest known Cambrian fossiliferous strata.

Further Discoveries

Since Darwin's time, the fossil record has been extended to between 2.3 and 3.5 billion years. Most of these Precambrian fossils are microscopic bacteria or microfossils. However, macroscopic fossils are now known from the late Proterozoic. The Ediacara biota (also called Vendian biota) dating from 575 million years ago collectively constitutes a richly diverse assembly of early multicellular eukaryotes.

The fossil record and faunal succession form the basis of the science of biostratigraphy or determining the age of rocks based on embedded fossils. For the first 150 years of geology, biostratigraphy and superposition were the only means for determining the relative age of rocks. The geologic time scale was developed based on the relative ages of rock strata as determined by the early paleontologists and stratigraphers.

Since the early years of the twentieth century, absolute dating methods, such as radiometric dating (including potassium/argon, argon/argon, uranium series, and, for very recent fossils, radiocarbon dating) have been used to verify the relative ages obtained by fossils and to provide absolute ages for many fossils. Radiometric dating has shown that the earliest known stromatolites are over 3.4 billion years old.

Modern view

The fossil record is life's evolutionary epic that unfolded over four billion years as environmental conditions and genetic potential interacted in accordance with natural selection.

The Virtual Fossil Museum: Paleontology has joined with evolutionary biology to share the interdisciplinary task of outlining the tree of life, which inevitably leads backwards in time to Precambrian microscopic life when cell structure and functions evolved. Earth's deep time in the Proterozoic and deeper still in the Archean is only "recounted by microscopic fossils and subtle chemical signals." Molecular biologists, using phylogenetics, can compare protein amino acid or nucleotide sequence homology (i.e., similarity) to evaluate taxonomy and evolutionary distances among organisms, with limited statistical confidence. The study of fossils, on the other hand, can more specifically pinpoint when and in what organism a mutation first appeared. Phylogenetics and paleontology work together in the clarification of science's still dim view of the appearance of life and its evolution.

Niles Eldredge's study of the *Phacops* trilobite genus supported the hypothesis that modifications to the arrangement of the trilobite's eye lenses proceeded by fits and starts over millions of years during the Devonian. Eldredge's interpretation of the *Phacops* fossil record was that the aftermaths of the lens changes, but not the rapidly occurring evolutionary process, were fossilized. This and other data led Stephen Jay Gould and Niles Eldredge to publish their seminal paper on punctuated equilibrium in 1971.

Example of Modern Development

Synchrotron X-ray tomographic analysis of early Cambrian bilaterian embryonic microfossils yielded new insights of metazoan evolution at its earliest stages. The tomography technique provides previously unattainable three-dimensional resolution at the limits of fossilization. Fossils of two enigmatic bilaterians, the worm-like *Markuelia* and a putative, primitive protostome, *Pseudooides*, provide a peek at germ layer embryonic development. These 543-million-year-old embryos support the emergence of some aspects of arthropod development earlier than previously thought in the late Proterozoic. The preserved embryos from China and Siberia underwent rapid diagenetic phosphatization resulting in exquisite preservation, including cell structures. This research is a notable example of how knowledge encoded by the fossil record continues to contribute otherwise unattainable information on the emergence and development of life on Earth. For example, the research suggests *Markuelia* has closest affinity to priapulid worms, and is adjacent to the evolutionary branching of Priapulida, Nematoda and Arthropoda.

Trading and Collecting

Fossil trading is the practice of buying and selling fossils. This is many times done illegally with artifacts stolen from research sites, costing many important scientific specimens each year. The problem is quite pronounced in China, where many specimens have been stolen.

Fossil collecting (some times, in a non-scientific sense, fossil hunting) is the collection of fossils for scientific study, hobby, or profit. Fossil collecting, as practiced by amateurs, is the predecessor of modern paleontology and many still collect fossils and study fossils as amateurs. Professionals and amateurs alike collect fossils for their scientific value.

Sporophyte

A sporophyte is the diploid multicellular stage in the life cycle of a plant or alga. It develops from the zygote produced when a haploid egg cell is fertilized by a haploid sperm and each cell therefore has a double set of chromosomes, one set from each parent. All land plants, and most multicellular algae, have life cycles in which a multicellular diploid sporophyte phase alternates with a multicellular haploid gametophyte phase. In the seed plants, the Gymnosperms and flowering plants (*Angiosperms*), the sporophyte phase is more prominent than the gametophyte, and is the familiar green plant with its roots, stem, leaves and cones or flowers. In flowering plants the gametophytes are very reduced in size, and are represented by the pollen and the embryo sac.

The sporophyte produces spores (hence the name), by meiosis. These meiospores develop into a gametophyte. Both the spores and the resulting gametophyte are haploid, meaning they only have one set of homologous chromosomes. The mature gametophyte produces male or female gametes (or both) by mitosis. The fusion of male and female gametes produces a diploid zygote which develops into a new sporophyte. This cycle is known as alternation of generations or alternation of phases.

Figure: *In flowering plants, the sporophyte comprises the whole multicellular body except the pollen and embryo sac*

Bryophytes (mosses, liverworts and hornworts) have a dominant gametophyte phase on which the adult sporophyte is dependent for nutrition. The embryo sporophyte develops by cell division of the zygote within the female sex organ or archegonium, and in its early development is therefore nurtured by the gametophyte. Because this embryo-nurturing feature of the life cycle is common to all land plants they are known collectively as the Embryophytes.

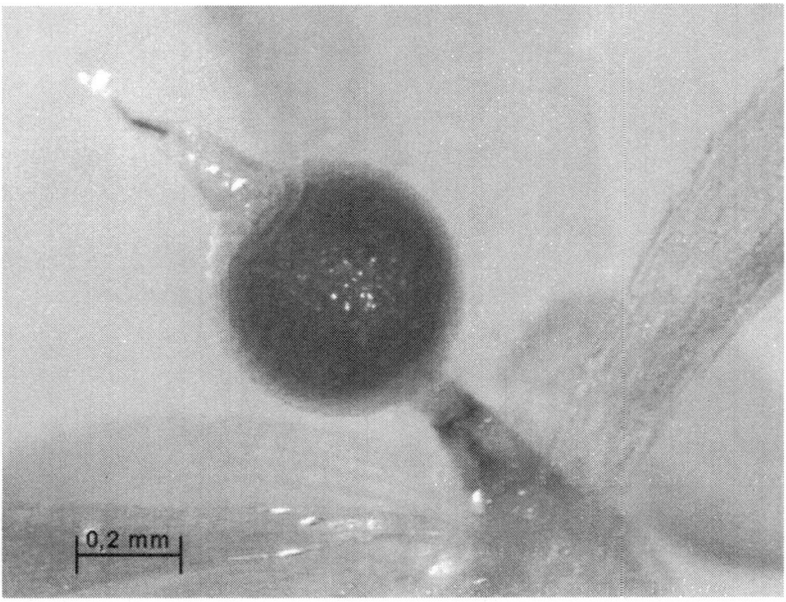

Figure: *Cleistocarpous sporophyte of the moss* Physcomitrella patens

Most algae have dominant gametophyte generations, but in some species the gametophytes and sporophytes are morphologically similar (isomorphic). An independent sporophyte is the dominant form in all clubmosses, horsetails, ferns, gymnosperms, and angiosperms (flowering plants) that have survived to the present day. Early land plants had sporophytes that produced identical spores (isosporous or homosporous) but the ancestors of the gymnosperms evolved complex heterosporous life cycles in which the spores producing male and female gametophytes were of different sizes, the female megaspores tending to be larger, and fewer in number, than the male microspores.

During the Devonian period several plant groups independently evolved heterospory and subsequently the habit of endospory, in which the gametophytes develop in miniaturized form inside the spore wall. By contrast in exosporous plants, including modern ferns, the gametophytes break the spore wall open on germination and develop outside it. The megagametophytes of endosporic plants such as the seed ferns developed within the sporangia of the parent sporophyte, producing a miniature multicellular female gametophyte complete with female sex organs, or archegonia. The oocytes were fertilized in the archegonia by free-swimming flagellate sperm produced by windborne miniatuarised male gametophytes in the form of pre-pollen. The resulting zygote developed into the next sporophyte generation while still retained within the pre-ovule, the single large female meiospore or megaspore contained in the modified sporangium or nucellus of the parent sporophyte. The evolution of heterospory and endospory were among the earliest steps in the evolution of seeds of the kind produced by gymnosperms and angiosperms today.

Psilotum

Psilotum (whisk fern) is a genus of fern-like vascular plants, one of two genera in the family Psilotaceae, order Psilotales, and class Psilotopsida (the other being *Tmesipteris*). The name of the genus is from Greek *psilos* = naked, because it lacks leaves that are normally found in other ferns.

Species and Distribution

There are two species, *Psilotum nudum* and *Psilotum complanatum*, with a hybrid between them known, *Psilotum* × *intermedium* W. H. Wagner.

The distribution of *Psilotum* is tropical and subtropical, in the New World, Asia, and the Pacific, with a few isolated populations in

SW Europe. The highest latitudes known are in South Carolina, Cádiz province in Spain, and southern Japan for *P. nudum*. In the U.S., one species is found from Florida to Texas, the other in Hawaii.

Relation to Ferns

Psilotum superficially resembles certain extinct early vascular plants, such as the rhyniophytes and the trimerophyte genus *Psilophyton*. The unusual features of *Psilotum* that suggest an affinity with early vascular plants include dichotomously branching sporophytes, aerial stems arising from horizontal rhizomes, a simple vascular cylinder, homosporous and terminal eusporangia and a lack of roots. Unfortunately, no fossils of psilophytes are known to exist. A careful study of the morphology and anatomy suggests that whisk ferns are not closely related to rhyniophytes, and that the ancestral features present in living psilophytes represent a reduction from a more typical modern fern plant.

Molecular evidence strongly confirms that *Psilotum* is a fern and that psilophytes are sister to ophioglossoid ferns.

Gametophyte

The gametophyte of *Psilotum nudum* L. Beauv. (*P. triquetrum* Swartz) is unusual in that it branches dichotomously, lives underground and possesses vascular tissue. The nutrition of the gametophyte appears to be myco-heterotrophic, assisted by endophytic fungi.

Psilotum Nudum

Psilotum nudum is a fernlike plant. Like the other species in the order Psilotales, it lacks roots. It is found in tropical Africa, Central America, tropical and subtropical North America, South America, tropical Asia, Australia, Hawaii, southern Japan, Lord Howe Island, New Zealand, with a few isolated populations in SW Europe ("Los Alcornocales", Spain, Cádiz province).

Psilotum nudum, known as the Skeleton Fork Fern is considered a "primitive" plant - a descendent of possibly the first group of vascular plants which were widespread during the Devonian and Silurian periods. Its name, *Psilotum nudum*, means "bare naked" in Latin, because it lacks most of the organs of modern plants.

In tropical areas, this plant is often epiphytic, whilst in more temperate areas, such as south-eastern Australia, it is usually found growing in rock crevices. Thousands of people per day walk by these plants at the Sydney Opera House forecourt.

Cultivation and Uses

The plant, which grows wild in southern Japan, was once much cultivated in Japanese gardens as an ornamental plant. Slightly over 100 garden varieties were given fantastic names. Called Matsubaran (pine needle orchid) in Japanese, it was one of the noble plants in the Edo Period (1603-1867).

The species is naturally found on all the main Hawaiian Islands. Known locally as Moa because of its chicken feet like stems, the Hawaiians collected large quantities of the spores and used them like talcum powder, under the loincloth to prevent chafing. The spores were also used medicinally as a purge. Children would play a game called 'moa nahele' or cock fighting with the branch stems. Twigs of the Moa were interlocked and the players pulled on the ends. The loser's twig broke and the winner crowed like a rooster. The species is still often used in making traditional Hawaiian leis.

Its common name, Whisk Fern, alludes to its use in the past as a small broom, made by tying a handful of its branches together. It is sometimes found in cultivation (either accidentally, as a weed in greenhouses, or deliberately, in the form of a number of cultivars).

It may prove to be a good source of antimicrobial chemicals.

Fungi Imperfecti

The Fungi imperfecti or imperfect fungi, also known as Deuteromycota, are fungi which do not fit into the commonly established taxonomic classifications of fungi that are based on biological species concepts or morphological characteristics of sexual structures because their sexual form of reproduction has never been observed; hence the name "imperfect fungi." Only their asexual form of reproduction is known, meaning that this group of fungi produces their spores asexually.

The *Deuteromycota* (Greek for "second fungi") were once considered a formal phylum of the kingdom Fungi. The term is now used only informally, to denote species of fungi that are asexually reproducing members of the fungal phyla Ascomycota and Basidiomycota.

There are about 25,000 species that have been classified in the deuteromycota. Fungi producing the antibiotic penicillin and those that cause athlete's foot and yeast infections are imperfect fungi. In addition, there are a number of edible imperfect fungi, including the ones that provide the distinctive characteristics of Roquefort and Camembert cheese.

Other, more informal, names besides Deuteromycota ("Deuteromycetes") and fungi imperfecti, are anamorphic fungi, or mitosporic fungi, but these are terms without taxonomic rank.

Problems in Taxonomic Classification

Although Fungi imperfecti/Deuteromycota is no longer formally accepted as a taxon, many of the fungi it included have yet to find a place in modern fungal classification. This is because most fungi are classified based on characteristics of the fruiting bodies and spores produced during sexual reproduction, and members of the Deutromycota have only been observed to produce asexual or no spores.

For this reason, mycologists are unique among those who study extant organisms in using a dual system of nomenclature. Dual naming was permitted by Article 59 of the International Code of Botanical Nomenclature (which governs the naming of plants and fungi); however, this was abolished in the 2011 update of the Code. Under the former system, a name for an asexually reproducing fungus was considered a *form taxon*. For example, the ubiquitous and industrially important mold, *Aspergillus niger*, has no known sexual cycle. Thus *Aspergillus niger* is considered a form taxon. In contrast, isolates of its close relative, *Aspergillus nidulans*, revealed it to be the anomorphic stage of a teleomorph already named *Emericella nidulans*. When a teleomorphic stage is known, that name will take priority over the name of an anamorph, hence this formerly classified *Aspergillus* species is now properly called *Emericella nidulans*.

Phylogeny and Taxonomy

Phylogenetic classification of asexually reproducing fungi now commonly uses molecular systematics. Phylogenetic trees constructed from comparative analyses of DNA sequences, such as RNA, or multigene phylogenies may be used to infer relationships between asexually reproducing fungi and their sexually reproducing counterparts. With these methods, many asexually reproducing fungi have now been placed in the tree of life. However, because phylogenetic methods require sufficient quantities of biological materials (spores or fresh specimens) that are from pure (i.e., uncontaminated) fungal cultures, for many asexual species their exact relationship with other fungal species has yet to be determined. Under the current system of fungal nomenclature, teleomorph names cannot be applied to fungi that lack sexual structures. Classifying and naming asexually reproducing fungi is the subject of ongoing debate in the mycological community.

Historical Classification of the Imperfect Fungi

These groups are no longer formally accepted because they do not adhere to the principle of monophyly. The taxon names are sometimes used informally. In particular, the term 'hyphomycetes' is often used to refer to molds, and the term 'coelomycetes' is used to refer to many asexually reproducing plant pathogens that form discrete fruiting bodies. Other systems of classification are reviewed by Kendrick (1981).

- Class Hyphomycetes lacking fruiting bodies
  - Order Moniliales (producing spores on simple conidiophores)
  - Order Stilbellales (producing spores on synnemata)
  - Order Tuberculariales (producing spores in sporodochia)
- Class Coelomycetes spores produced in fruiting bodies
  - Order Melanconiales (producing spores in acervuli)
  - Order Sphaeropsidales (producing spores in pycnidia)
- Class Agonomycetes lacking spores

Common Species

Industrially relevant fungi:

- *Tolypocladium inflatum* '! from which we obtain the immunosuppressant ciclosporin;
- *Penicillium chrysogenum*
- *Penicillium griseofulvum*
- *Penicillium roqueforti*
- *Penicillium camemberti*
- Other species of *Penicillium* are used to improve both the taste and the texture of cheeses
- *Aspergillus oryzae*
- *Aspergillus sojae*
- *Aspergillus niger*
- *Cladosporium resinae*
- *Lecanicillium* sp. '! these produce conidia which may control certain species of insect pests
- Other entomopathogenic fungi, including *Metarhizium* and *Beauveria* spp.
- *Pochonia* spp. are under development for control of Nematode pests.

Lepidodendron

Lepidodendron — known as scale trees — were a now extinct genus of primitive, vascular, arborescent (tree-like) plant related to the lycopsids (club mosses). They were part of the coal forest flora. They sometimes reached heights of over 30 metres (100 ft), and the trunks were often over 1 m (3.3 ft) in diameter, and thrived during the Carboniferous Period (about 359.2 ± 2.5 Mya (million years ago) to about 299.0 ± 0.8 Mya). Sometimes erroneously called "giant club mosses", they were actually more closely related to today's quillworts than to modern club mosses.

The name *Lepidodendron* comes from the Greek *lepido*, scale, and *dendron*, tree.

Description and Biology

Lepidodendron had tall, thick trunks that rarely branched and were topped with a crown of bifurcating branches bearing clusters of leaves. These leaves were long and narrow, similar to large blades of grass, and were spirally-arranged. The vascular system was a siphonostele with exarch xylem maturation.

The closely packed diamond-shaped leaf scars left on the trunk and stems as the plant grew provide some of the most interesting and common fossils in Carboniferous shales and accompanying coal deposits. These fossils look much like tire tracks or alligator skin.

The scars, or leaf cushions, were composed of green photosynthetic tissue, evidenced by the cuticle covering and being dotted with stomata, microscopic pores through which carbon dioxide from the air diffuses into plants. Likewise, the trunks of *Lepidodendron* would have been green, unlike modern trees which have scaly, non-photosynthetic brown or gray bark.

Lepidodendron has been likened to a giant herb. The trunks produced little wood, being mostly soft tissues. Most structural support came from a thick, bark-like region. This region remained around the trunk as a rigid layer that grew thicker, but did not flake off like that of most modern trees. As the tree grew, the leaf cushions expanded to accommodate the increasing width of the trunk.

Lepidodendron likely lived in the wettest parts of the coal swamps that existed during the Carboniferous period. They grew in dense stands, likely having as many as 1000 to 2000 giant clubmosses per hectare. This would have been possible because they did not branch until fully grown, and would have spent much of their lives as unbranched poles.

Reproduction

The branches of this plant ended in cone-structures. *Lepidodendron* did not produce true seeds like many modern plants. Instead, it reproduced by means of elaborate, encapsulated spores. It is estimated that these plants grew rapidly and lived 10–15 years. Most species were probably monocarpic, meaning they reproduced only once toward the end of their life cycle.

Decline and Extinction

By the Mesozoic era, the giant lycopsids had died out and were replaced by smaller Quillworts, probably due to competition from the emerging woody gymnosperms and other plants. *Lepidodendron* is one of the more common plant fossils found in Pennsylvanian (Late Carboniferous) age rocks. They are closely related to other extinct Lycopsid genera, *Sigillaria* and *Lepidendropsis.*

Rhyniopsida

Rhyniopsida is a class of extinct early vascular plants, with one family, Rhyniaceae, found in the Early Devonian (around 419 to 393 million years ago). They are polysporangiophytes, since their sporophytes consisted of branched stems bearing sporangia (spore-forming organs). They lacked leaves or true roots but did have simple vascular tissue.

The group was first placed in a subdivision of the division Tracheophyta under the name Rhyniophytina. Informally, they are often called rhyniophytes. More recently the name paratracheophytes has been suggested, to distinguish such plants from 'true' tracheophytes or eutracheophytes.

As originally defined, the group was found not to be monophyletic since some of its members are now known to lack vascular tissue; these have been moved to the class Horneophytopsida, which is defined as lacking true vascular tissue. Currently, Rhyniopsida includes the genera *Huvenia*, *Rhynia*, and *Stockmansella*, all from the Devonian.

One of the most important radiations for land plants occurred in the early Devonian (Pragian), when the first certain rhyniophytes appear in the fossil record, making this rich fossil discovery of major importance to paleobotany. It has been suggested that the poorly preserved *Eohostimella*, found in deposits of Early Silurian age (Llandovery, around 440 to 430 million years ago), may be a rhyniophyte.

gametophytes. Moss gametophytes originate from the germination of a spore. The initial phase of growth leads to a filament of cells (called the protonema). The mature gametophyte of mosses produces the gamete-producing structures, archegonia and antheridia. In gametophyte dominant organisms, such as bryophytes, specialized structures or even whole individual plants that produce gametes are sometimes called gametophores.

In other living land plants (the vascular plants), an interesting pattern exists with regard to gametophytes. Homosporous vascular plants have exosporic gametophytes—that is, the gametophyte develops outside of the spore wall. In addition, these exosporic gametophytes are normally bisexual (produce both sperm and eggs). In heterosporous vascular plants, the gametophyte develops endosporically—inside the wall of the spore. These gametophytes are unisex, producing either sperm or eggs but not both. All vascular plants are sporophyte dominant, and a trend toward smaller and shorter-lived gametophytes is evident as land plants became increasingly adapted to a terrestrial environment.

In modern clubmosses the gametophyte is subterranean and mycotrophic, deriving its nutrients from symbiotic fungi. In lycophytes, which are heterosporous, the megagametophyte develops inside the spore, which cracks open to allow access to internal archegonia.

In many ferns the gametophyte is a photosynthetic free living organism called a prothallus. However in some groups, notably Ophioglossaceae and *Psilotum*, the gametophytes are subterranean and subsist by forming mycotrophic relationships with fungi. In leptosporangiate ferns such as *Dryopteris* the gametophyte is a free-living, autotrophic prothallium that maintains the sporophyte during its early multicellular development. By contrast, in seed plants (gymnosperms and angiosperms) the gametophytes are nurtured by the sporophyte during development and become multicellular while still enclosed within the sporangium.

Phylogeny

In 2004, Crane et al. published a cladogram for the polysporangiophytes in which the Rhyniaceae are shown as the sister group of all other tracheophytes (vascular plants). The other former "rhyniophytes", such as *Horneophyton* and *Aglaophyton*, are placed outside the tracheophyte clade, as they did not possess true vascular tissue (in particular did not have tracheids).

Rhynie Flora

The general term "rhyniophytes" or "rhyniophytoids" is sometimes used for the assemblage of plants found in the Rhynie chert Lagerstätte - rich fossil beds in Aberdeenshire, Scotland, and roughly coeval sites with similar flora. Used in this way, these terms refer to a floristic assemblage of more or less related early land plants, not a taxon. Though the rhyniophytes are well represented, plants with simpler anatomy, like *Aglaophyton*, are also common; there are also more complex plants, like *Asteroxylon*, which has a very early form of leaves.

The Rhynie flora is unusual for the excellent preservation of early vascular plants, in addition to plants transitional between vascular and non-vascular. The fossils contain sufficient internal detail to determine vascular organization and to distinguish sporangia and gametangia. This has led to the recognition of species which apparently had an isomorphic alternation of generations (gametophytes and sporophytes of similar prominence), a condition unknown among land plants today. Because the plants were buried in situ, rather than after transport to a distant location, important morphological details and ecological information can be obtained. The site also preserves other organisms such as arthropods and fungi that lived in the Rhynie ecosystem.

Gametophyte

In heterosporous plants (ferns, lycophytes, angiosperms and gymnosperms) the egg producing gametophyte is known as a megagametophyte and the sperm producing gametophyte as a microgametophyte. In the seed plants, the immature microgametophyte is called pollen. Seed plant microgametophytes consists of two or three cells when the pollen grains exit the sporangium. The megagametophyte develops within the megaspore of extant seedless vascular plants and within the megasporangium in a cone or flower in seed plants. In seed plants, the microgametophyte (pollen grain) travels to the vicinity of the egg cell (carried by a physical or animal vector), and produces two sperm by mitosis.

In gymnosperms the megagametophyte consists of several thousand cells and produces one to several archegonia, each with a single egg cell. The gametophyte becomes a food storage tissue in the seed. as the seed germinates.

In angiosperms, the megagametophyte is reduced to only a few nuclei and cells, and is sometimes called the embryo sac. A typical flowering plant ovule contains seven cells and eight nuclei, one of which is the egg cell. Two nuclei fuse with a sperm nucleus form endosperm, which becomes the food storage tissue in the seed.

In some multicellular green algae, red algae, or brown algae (*Ulva* is one example) sporophytes and gametophytes may be externally indistinguishable or the gametophyte may be reduced.

6

Synonymous Substitution

A synonymous substitution (also called a *silent* substitution) is the evolutionary substitution of one base for another in an exon of a gene coding for a protein, such that the produced amino acid sequence is not modified. Synonymous substitutions and mutations affecting noncoding DNA are collectively known as silent mutations. A nonsynonymous substitution results in a change in amino acid that may be arbitrarily further classified as conservative (change to an amino acid with similar physiochemical properties), semi-conservative (e.g. negative to positively charged amino acid), or radical (vastly different amino acid).

Degeneracy of the Genetic Code

Protein translation involves a set of twenty amino acids. Each of these amino acids is coded for by a sequence of three DNA base pairs called a *codon*. Because there are 64 possible codons, but only 20 amino acids (as well as a stop signal [i.e. the three codons that do not code for any amino acid and are known as stop codons], indicating that translation should stop), some amino acids are coded for by 2, 3, 4, or 6 different codons. For example, the codons TTT and TTC both code for the amino acid phenylalanine. This is often referred to as *redundancy* of the genetic code. There are two mechanisms for redundancy: several different transfer RNAs can deliver the same amino acid, or one tRNA can have a non-standard "wobbly" base in position three of the anti-codon, which recognises more than one base in the codon.

In the above phenylalanine example, suppose that the base in position 3 of a TTT codon got substituted to a C, leaving the codon

TTC. The amino acid at that position in the protein will remain a phenylalanine. Hence, the substitution is a synonymous one.

Substitution Versus Mutation

Although mutation and substitution are often used interchangeably, there is a subtle but important difference. A nucleotide mutation is a base change (whether synonymous or non-synonymous) such that the mutant and wild-type forms coexist in a population. A nucleotide substitution is a base change between two populations. Thus, a mutation only becomes a substitution when the most recent common ancestor of the entire population carried that mutation. When all lineages carrying alternative mutations have died off, the remaining mutation is said to be *fixed*. Note that fixed mutations may never reach 100% frequency in the population, as further mutations at the same site may arise; these subsequent mutations, however, will all share a common ancestor which had the fixed mutation.

Synonymous Substitutions and Evolution

When a synonymous or silent mutation occurs, the change is often assumed to be neutral, meaning that it does not affect the fitness of the individual carrying the new gene to survive and reproduce.

Synonymous changes may not be neutral because certain codons are translated more efficiently (faster and/or more accurately) than others. For example, when a handful of synonymous changes in the fruit fly alcohol dehydrogenase gene were introduced, changing several codons to sub-optimal synonyms, production of the encoded enzyme was reduced and the adult flies showed lower ethanol tolerance. Many organisms, from bacteria through animals, display biased use of certain synonymous codons. Such codon usage bias may arise for different reasons, some selective, and some neutral. In *Saccharomyces cerevisiae* synonymous codon usage has been shown to influence mRNA folding stability, with mRNA encoding different protein secondary structure preferring different codons.

Another reason why synonymous changes are not always neutral is the fact that exon sequences close to exon-intron borders function as RNA splicing signals. When the splicing signal is destroyed by a synonymous mutation, the exon does not appear in the final protein. This results in a truncated protein. One study found that about a quarter of synonymous variations affecting exon 12 of the cystic ûbrosis transmembrane conductance regulator gene result in that exon being skipped.

Point Mutation

A point mutation, or single base substitution, is a type of mutation that causes the replacement of a single base nucleotide with another nucleotide of the genetic material, DNA or RNA. The term *point mutation* also includes insertions or deletions of a single base pair.

A point mutant is an individual that is affected by a point mutation.

Repeat induced point mutations are recurring point mutations, discussed below.

Causes of Point Mutations

Point mutation is a random SNP (single nucleotide polymorphism) mutation in the deoxyribonucleic acid (DNA) that occurs at one point. Point mutations usually take place during DNA replication. DNA replication occurs when one double-stranded DNA molecule creates two single strands of DNA that is a template for the creation of the coinciding strand. A single point mutation can change the whole DNA sequence. Changing one purine or pyrimidine may change the amino acid that the nucleotides code for.

Point mutations may arise from spontaneous mutations that occur during DNA replication. The rate of mutation may be increased by mutagens. Mutagens can be physical, such as radiation from UV rays, X-rays or extreme heat, or chemical (molecules that misplace base pairs or disrupt the helical shape of DNA). Mutagens associated with cancers are often studied to learn about cancer and its prevention.

There are multiple ways for point mutations to occur. First, ultraviolet(UV) light and higher-frequency light are capable ionizing electrons and in turn impacting DNA. Also, one of the cell metabolic byproducts, reactive oxygen molecules with free radicals, can also be very harmful to DNA. These reactants can lead to both single-stranded DNA breaks and double-stranded DNA breaks. Third, bonds in DNA eventually degrade, which creates another problem to keep the integrity of DNA to a high standard. There can also be replication errors that lead to substitution, insertion, or deletion mutations.

It was previously believed that these mutations happened completely by chance, with no regard for their effects on the organisms. Recently, there have been studies suggesting that these mutations occur in response to environmental challenges. That is to say, they are more likely to occur when they are advantageous to the organism, rather than when they are neutral or disadvantageous. When cells were deprived of a certain amino acid, tryptophan, for prolonged periods of time, point mutations in trp operon reverted to tryptophan, leading to an advantageous result, more frequently than under normal conditions when the mutations were neutral.

In addition, the tryptophan mutation rate was unaffected when the cells were deprived of another amino acid, cysteine, further suggesting that the mutation rate was specific to situations in which the mutation was advantageous.

Categorizing Point Mutations

Transition/Transversion Categorization: In 1959 Ernst Freese coined the terms "transitions" or "transversions" to categorize different types of point mutations.

- Transitions: replacement of a purine base with another purine or replacement of a pyrimidine with another pyrimidine
- Transversions: replacement of a purine with a pyrimidine or vice versa.

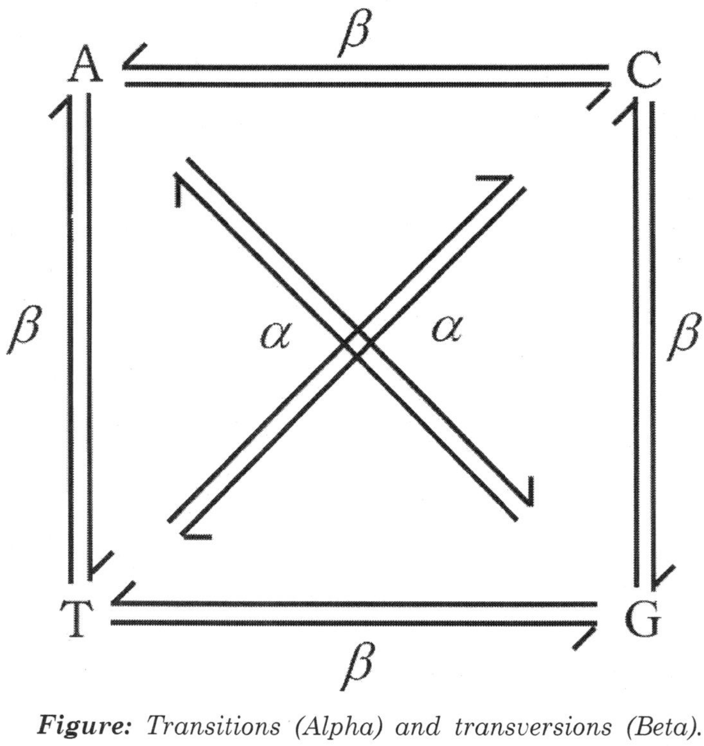

Figure: *Transitions (Alpha) and transversions (Beta).*

There is a systematic difference in mutation rates for transitions (Alpha) and transversions (Beta). Transition mutations are about an order of magnitude more common than transversions.

Functional Categorization

- Nonsense mutations: Code for a stop, which can truncate the protein. A nonsense mutation converts an amino acid codon into a termination codon. This causes the protein to be shortened because of the stop codon interrupting its normal code. How much of the protein is lost determines whether or not the protein is still functional.
- Missense mutations: Code for a different amino acid. A missense mutation changes a codon so that a different protein is created, a non-synonymous change.
    - o Conservative mutations: Result in an amino acid change. However, the properties of the amino acid remain the same (e.g., hydrophobic, hydrophilic, etc.). At times, a change to one Amino acid in the protein is not detrimental to the organism as a whole. Most proteins can withstand one or two point mutations before their function changes.
    - o Non-conservative mutations: Result in an amino acid change that has different properties than the wild type. The protein may lose its function, which can result in a disease in the organism. For example, sickle-cell disease is caused by a single point mutation (a missense mutation) in the beta-hemoglobin gene that converts a GAG codon into GUG, which encodes the amino acid valine rather than glutamic acid. The protein may also exhibit a "gain of function" or become activated, such is the case with the mutation changing a valine to glutamic acid in the braf gene; this leads to an activation of the RAF protein which causes unlimited proliferative signalling in cancer cells. These are both examples of a non-conservative (missense) mutation.
- Silent mutations: Code for the same amino acid. A silent mutation has no effect on the functioning of the protein. A single nucleotide can change, but the new codon specifies the same amino acid, resulting in an unmutated protein. This type of change is called synonymous change, since the old and new codon code for the same amino acid. This is possible because 64 codons specify only 20 amino acids. Different codons can lead to differential protein expression levels, however.

Use of Term to Describe Single Base Pair Insertions and Deletions

- Sometimes the term *point mutation* is used to describe insertions or deletions of a single base pair (which has more of an adverse effect on the synthesized protein due to the nucleotides' still being read in triplets, but in different frames: a mutation called a frameshift mutation).

General Consequences of Point Mutations

Point mutations that occur in non-coding sequences are most often without consequences, although there are exceptions. If the mutated base pair is in the promoter sequence of a gene, then the expression of the gene may change. Also, if the mutation occurs in the splicing site of an intron, then this may interfere with correct splicing of the transcribed pre-mRNA.

By altering just one amino acid, the entire peptide may change, thereby changing the entire protein. If the protein functions in cellular reproduction then this single point mutation can change the entire process of cellular reproduction for this organism.

Point germline mutations can lead to beneficial as well as harmful traits or diseases. This leads to adaptations based on the environment where the organism lives. An advantageous mutation can create an advantage for that organism and lead to the trait's being passed down from generation to generation, improving and benefiting the entire population. The scientific theory of evolution is greatly dependent on point mutations in cells. The theory explains the diversity and history of living organisms on Earth. In relation to point mutations, it states that beneficial mutations allow the organism to thrive and reproduce, thereby passing its positively affected mutated genes on to the next generation. On the other hand, harmful mutations cause the organism to die or be less likely to reproduce in a phenomenon known as natural selection.

There are different short-term and long-term effects that can arise from mutations. Smaller ones would be a halting of the cell cycle at numerous points. This means that a codon coding for the amino acid glycine may be changed to a stop codon, causing the proteins that should have been produced to be deformed and unable to complete their intended tasks. Because the mutations can affect the DNA and thus the chromatin, it can prohibit mitosis from occurring due to the lack of a complete chromosome. Problems can also arise during the processes of transcription and replication of DNA. These all prohibit the cell from reproduction and thus lead to the death of the cell. Long-term effects can be a permanent changing of a chromosome, which can lead to a mutation. These mutations can be either beneficial or detrimental. Cancer is an example of how they can be detrimental.

Other effects of point mutations, or single nucleotide polymorphisms in DNA, depend on the location of the mutation within the gene. For example, if the mutation occurs in the region of the gene responsible for coding, the amino acid sequence of the encoded protein may be altered, causing a change in the function, activation, localization, or stability of the protein. Moreover, if the mutation occurs in the region of the gene where transcriptional machinery binds to the protein, the mutation can affect the binding of the transcription factors because the short nucleotide sequences recognized by the transcription factors will be altered. Mutations in this region can affect rate of efficiency of gene transcription, which in turn can alter levels of mRNA and, thus, protein levels in general.

Point mutations can have several effects on the behaviour and reproduction of a protein depending on where the mutation occurs in the amino acid sequence of the protein. If the mutation occurs in the region of the gene that is responsible for coding for the protein, the amino acid may be altered. This slight change in the sequence of amino acids can cause a change in the function, activation of the protein meaning how it binds with a given enzyme, where the protein will be located within the cell, or the amount of free energy stored within the protein.

If the mutation occurs in the region of the gene where transcriptional machinery binds to the protein, the mutation can affect the way in which transcription factors bind to the protein. The mechanisms of transcription bind to a protein through recognition of short nucleotide sequences. A mutation in this region may alter these sequences and, thus, change the way the transcription factors bind to the protein. Mutations in this region can affect the efficiency of gene transcription, which controls both the levels of mRNA and overall protein levels.

Specific Diseases Caused by Point Mutations

Cystic Fibrosis: A defect in the cystic fibrosis transmembrane conductance regulator (CFTR) gene causes cystic fibrosis (CF). A protein made by this gene controls the movement of the water and salt in and out of the body's cells. Genes in people with CF incorrectly code proteins. This causes thick, sticky mucus and very salty sweat.

Cancer

Point mutations in multiple tumor suppressor proteins cause cancer. For instance, point mutations in Adenomatous Polyposis Coli promote tumorigenesis. A novel assay, Fast parallel proteolysis (FASTpp), might help swift screening of specific stability defects in individual cancer patients.

Neurofibromatosis

Neurofibromatosis is caused by point mutations in the Neurofibromin 1 or Neurofibromin 2 gene.

Sickle-cell Anemia

Sickle-cell anemia is caused by a point mutation in the â-globin chain of haemoglobin, causing the hydrophilic amino acid glutamic acid to be replaced with the hydrophobic amino acid valine at the sixth position.

The â-globin gene is found on the short arm of chromosome 11. The association of two wild-type á-globin subunits with two mutant â-globin subunits forms haemoglobin S (HbS). Under low-oxygen conditions (being at high altitude, for example), the absence of a polar amino acid at position six of the â-globin chain promotes the non-covalent polymerisation (aggregation) of haemoglobin, which distorts red blood cells into a sickle shape and decreases their elasticity.

Hemoglobin is a protein found in red blood cells, and is responsible for the transportation of oxygen through the body. There are two subunits that make up the hemoglobin protein: beta-globins and alpha-globins. Beta-hemoglobin is created from the genetic information on the HBB, or "hemoglobin, beta" gene found on chromosome 11p15.5. A single point mutation in this polypeptide chain, which is 147 amino acids long, results in the disease known as Sickle Cell Anemia. Sickle-Cell Anemia is an autosomal recessive disorder that affects 1 in 500 African Americans, and is one of the most common blood disorders in the United States. The single replacement of the sixth amino acid in the beta-globin, glutamic acid, with valine results in deformed red blood cells. These sickle-shaped cells cannot carry nearly as much oxygen as normal red blood cells and they get caught more easily in the capillaries, cutting off blood supply to vital organs. The single nucleotide change in the beta-globin means that even the smallest of exertions on the part of the carrier results in severe pain and even heart attack. Below is a chart depicting the first thirteen amino acids in the normal and abnormal sickle cell polypeptide chain.

Sequence for Normal Hemoglobin

ATG	GTG	CAC	CTG	ACT	CCT	GAG	GAG	AAG	TCT	GCC	GTT	ACT
START	Val	His	Leu	Thr	Pro	Glu	Glu	Lys	Ser	Ala	Val	Thr

Sequence for Sickle Cell Hemoglobin

ATG	GTG	CAC	CTG	ACT	CCT	GtG	GAG	AAG	TCT	GCC	GTT	ACT
START	Val	His	Leu	Thr	Pro	Val	Glu	Lys	Ser	Ala	Val	Thr

Tay–Sachs Disease

The cause of Tay–Sachs disease is a genetic defect that is passed from parent to child. This genetic defect is located in the HEXA gene, which is found on chromosome 15.

The HEXA gene makes part of an enzyme called beta-hexosaminidase A, which plays a critical role in the nervous system. This enzyme helps break down a fatty substance called GM2 ganglioside in nerve cells. Mutations in the HEXA gene disrupt the activity of beta-hexosaminidase A, preventing the breakdown of the fatty substances. As a result, the fatty substances accumulate to deadly levels in the brain and spinal cord. The buildup of GM2 ganglioside causes progressive damage to the nerve cells. This is the cause of the signs and symptoms of Tay-Sachs disease.

Colour Blindness

People who are colourblind have mutations in their genes that cause a loss of either red or green cones, and they therefore have a hard time distinguishing between colours. There are three kinds of cones in the human eye: red, green, and blue. Now researchers have discovered that some people with the gene mutation that causes colourblindness lose an entire set of "colour" cones with no change to the clearness of their vision overall.

Repeat Induced Point-mutation

In molecular biology, repeat induced point-mutation or RIP is a process by which DNA accumulates G:C to A:T transition mutations. Genomic evidence indicates that RIP occurs or has occurred in a variety of fungi while experimental evidence indicates that RIP is active in *Neurospora crassa, Podospora anserina, Magnaporthe grisea, Leptosphaeria maculans, Gibberella zeae* and *Nectria haematococca.* In *Neurospora crassa,* sequences mutated by RIP are often methylated *de novo.*

RIP occurs during the sexual stage in haploid nuclei after fertilization but prior to meiotic DNA replication. In *Neurospora crassa,* repeat sequences of at least 400 base pairs in length are vulnerable to RIP. Repeats with as low as 80% nucleotide identity may also be subject to RIP. Though the exact mechanism of repeat recognition and mutagenesis are poorly understood, RIP results in repeated sequences undergoing multiple transition mutations.

The RIP mutations do not seem to be limited to repeated sequences. Indeed, for example, in the phytopathogenic fungus *L. maculans*, RIP mutations are found in single copy regions, adjacent to the repeated elements. These regions are either non-coding regions or genes encoding small secreted proteins including avirulence genes. The degree of RIP within these single copy regions was proportional to their proximity to repetitive elements.

Rep and Kistler have speculated that the presence of highly repetitive regions containing transposons, may promote mutation of resident effector genes. So the presence of effector genes within such regions is suggested to promote their adaptation and diversification when exposed to strong selection pressure.

As RIP mutation is traditionally observed to be restricted to repetitive regions and not single copy regions, Fudal *et al.* suggested that leakage of RIP mutation might occur within a relatively short distance of a RIP-affected repeat. Indeed, this has been reported in *N. crassa* whereby leakage of RIP was detected in single copy sequences at least 930 bp from the boundary of neighbouring duplicated sequences. To elucidate the mechanism of detection of repeated sequences leading to RIP may allow to understand how the flanking sequences may also be affected.

Mechanism

RIP causes G:C to A:T transition mutations within repeats, however, the mechanism that detects the repeated sequences is unknown. RID is the only known protein essential for RIP. It is a DNA methyltransferease-like protein, that when mutated or knocked out results in loss of RIP. Deletion of the *rid* homolog in *Aspergillus nidulans, dmtA*, results in loss of fertility while deletion of the *rid* homolog in *Ascobolus immersens, masc1*, results in fertility defects and loss of methylation induced premeiotically (MIP).

Consequences

RIP is believed to have evolved as a defence mechanism against Transposable Elements, which resemble parasites by invading and multiplying within the genome. RIP creates multiple missense and nonsense mutations in the coding sequence. This hypermutation of G-C to A-T in repetitive sequences eliminates functional gene products of the sequence (if there were any to begin with). In addition, many of the C-bearing nucleotides become methylated, thus decreasing transcription.

Use in Molecular Biology

Because RIP is so efficient at detecting and mutating repeats, fungal biologists often use it as a tool for mutagenesis. A second copy of a single-copy gene is first transformed into the genome. The fungus must then mate and go through its sexual cycle to activate the RIP machinery. Many different mutations within the duplicated gene are obtained from even a single fertilization event so that inactivated alleles, usually due to nonsense mutations, as well as alleles containing missense mutations can be obtained.

History

The Cellular Reproduction process of Meiosis was discovered by Oscar Hertwig in 1876. Mitosis was discovered several years later in 1882 by Walther Flemming.

Hertwig studied sea urchins, and noticed that each egg contained one nucleus prior to fertilization and two nuclei after. This discovery proved that one spermatozoon could fertilize an egg, and therefore proved the process of meiosis. Hermann Fol continued Hertwig's research by testing the effects of injecting several spermatozoa into an egg, and found that the process did not work with more than one spermatozoon.

Flemming began his research of cell division starting in 1868. The study of cells was an increasingly popular topic in this time period. By 1873, Schneider had already begun to describe the steps of cell division. Flemming furthered this description in 1874 and 1875 as he explained the steps in more detail. He also argued with Schneider's findings that the nucleus separated into rod-like structures by suggesting that the nucleus actually separated into threads that in turn separated. Flemming concluded that cells replicate through cell division, to be more specific mitosis.

Matthew Meselson and Franklin Stahl are credited with the discovery of DNA replication. Watson and Crick acknowledged that the structure of DNA did indicate that there is some form of replicating process. However, there was not a lot of research done on this aspect of DNA until after Watson and Crick. People considered all possible methods of determining the replication process of DNA, but none were successful until Meselson and Stahl. Meselson and Stahl introduced a heavy isotope into some DNA and traced its distribution. Through this experiment, Meselson and Stahl were able to prove that DNA reproduces semi-conservatively.

Plastid

The plastid is a major organelle found in the cells of plants and algae. Plastids are the site of manufacture and storage of important chemical compounds used by the cell. They often contain pigments used in photosynthesis, and the types of pigments present can change or determine the cell's colour. They possess a double-stranded DNA molecule, which is circular, like that of prokaryotes.

Plastids in Plants

Those plastids which contain pigments can carry out photosynthesis. Plastids can also store products like starch and can synthesise fatty acids and terpenes, which can be used for producing energy and as raw material for the synthesis of other molecules. For example, the components of the plant cuticle and its epicuticular wax, are synthesized by the epidermal cells from palmitic acid, which is synthesized in the chloroplasts of the mesophyll tissue. All plastids are derived from proplastids which are present in the meristematic regions of the plant. Proplastids and young chloroplasts commonly divide by binary fission, but more mature chloroplasts also have this capacity.

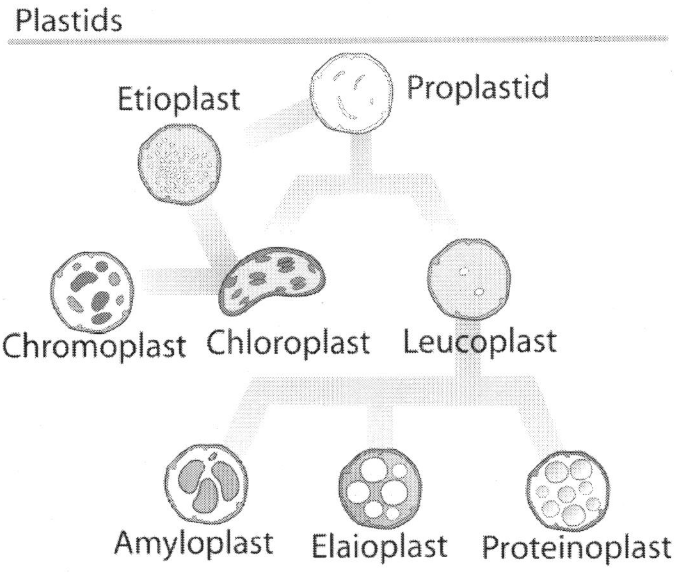

Figure: *In plants, plastids may differentiate into several forms, depending upon which function they play in the cell. Undifferentiated plastids* (proplastids) *may develop into any of the following variants:*

- Chloroplasts green plastids: for photosynthesis
- Chromoplasts coloured plastids: for pigment synthesis and storage
- Gerontoplasts: control the dismantling of the photosynthetic apparatus during senescence
- Leucoplasts colourless plastids: for monoterpene synthesis; *leucoplasts sometimes differentiate into more specialized plastids:*
    - o Amyloplasts: for starch storage and detecting gravity
    - o Elaioplasts: for storing fat
    - o Proteinoplasts: for storing and modifying protein

Depending on their morphology and function, plastids have the ability to differentiate, or redifferentiate, between these and other forms.

Each plastid creates multiple copies of a circular 75–250 kilobase plastome. The number of genome copies per plastid is variable, ranging from more than 1000 in rapidly dividing cells, which, in general, contain few plastids, to 100 or fewer in mature cells, where plastid divisions have given rise to a large number of plastids. The plastome contains about 100 genes encoding ribosomal and transfer ribonucleic acids (rRNAs and tRNAs) as well as proteins involved in photosynthesis and plastid gene transcription and translation. However, these proteins only represent a small fraction of the total protein set-up necessary to build and maintain the structure and function of a particular type of plastid. Plant nuclear genes encode the vast majority of plastid proteins, and the expression of plastid genes and nuclear genes is tightly co-regulated to coordinate proper development of plastids in relation to cell differentiation.

Plastid DNA exists as large protein-DNA complexes associated with the inner envelope membrane and called 'plastid nucleoids'. Each nucleoid particle may contain more than 10 copies of the plastid DNA. The proplastid contains a single nucleoid located in the centre of the plastid. The developing plastid has many nucleoids, localized at the periphery of the plastid, bound to the inner envelope membrane. During the development of proplastids to chloroplasts, and when plastids convert from one type to another, nucleoids change in morphology, size and location within the organelle. The remodelling of nucleoids is believed to occur by modifications to the composition and abundance of nucleoid proteins.

Many plastids, particularly those responsible for photosynthesis, possess numerous internal membrane layers.

In plant cells, long thin protuberances called stromules sometimes form and extend from the main plastid body into the cytosol and interconnect several plastids. Proteins, and presumably smaller molecules, can move within stromules. Most cultured cells that are relatively large compared to other plant cells have very long and abundant stromules that extend to the cell periphery.

Plastids in Algae

In algae, the term leucoplast is used for all unpigmented plastids and their function differs from the leucoplasts of plants. Etioplasts, amyloplasts and chromoplasts are plant-specific and do not occur in algae. Plastids in algae and hornworts may also differ from plant plastids in that they contain pyrenoids.

Glaucocystophytic algae contain muroplasts, which are similar to chloroplasts except that they have a cell wall that is similar to that of prokaryotes. Rhydophytic algae contain rhydoplasts, which are red chloroplasts that allow the algae to photosynthesise to a depth of up to 268 m.

Inheritance of Plastids

Most plants inherit the plastids from only one parent. In general, angiosperms inherit plastids from the female gamete, whereas many gymnosperms inherit plastids from the male pollen. Algae also inherit plastids from only one parent. The plastid DNA of the other parent is, thus, completely lost.

In normal intraspecific crossings (resulting in normal hybrids of one species), the inheritance of plastid DNA appears to be quite strictly 100% uniparental. In interspecific hybridisations, however, the inheritance of plastids appears to be more erratic. Although plastids inherit mainly maternally in interspecific hybridisations, there are many reports of hybrids of flowering plants that contain plastids of the father. Approximately 20% of angiosperms, including alfalfa (*Medicago sativa*), normally show biparental inheritance of plastids.

Origin of Plastids

Plastids are thought to have originated from endosymbiotic cyanobacteria. The symbiosis evolved around 1500 million years ago and enabled eukaryotes to carry out oxygenic photosynthesis. Three evolutionary lineages have since emerged in which the plastids are named differently: chloroplasts in green algae and plants, rhodoplasts in red algae and muroplasts in the glaucophytes. The plastids differ by their pigmentation, but also in ultrastructure. The chloroplasts, e.g., have lost all phycobilisomes, the light harvesting complexes found in cyanobacteria, red algae and glaucophytes, but instead contain stroma and grana thylakoids, structures found only in plants and in closely related green algae. The glaucocystophycean plastid — in contrast to the chloroplasts and the rhodoplasts — is still surrounded by the remains of the cyanobacterial cell wall. All these primary plastids are surrounded by two membranes.

Complex plastids start by secondary endosymbiosis, when a eukaryote engulfs a red or green alga and retains the algal plastid, which is typically surrounded by more than two membranes. In some cases these plastids may be reduced in their metabolic and/or photosynthetic capacity. Algae with complex plastids derived by secondary endosymbiosis of a red alga include the heterokonts, haptophytes, cryptomonads, and most dinoflagellates (= rhodoplasts). Those that endosymbiosed a green alga include the euglenids and chlorarachniophytes (= chloroplasts).

The Apicomplexa, a phylum of obligate parasitic protozoa including the causative agents of malaria (*Plasmodium* spp.), toxoplasmosis (*Toxoplasma gondii*), and many other human or animal diseases also harbor a complex plastid (although this organelle has been lost in some apicomplexans, such as *Cryptosporidium parvum*, which causes cryptosporidiosis). The 'apicoplast' is no longer capable of photosynthesis, but is an essential organelle, and a promising target for antiparasitic drug development.

Some dinoflagellates and sea slugs, in particular of the genus Elysia, take up algae as food and keep the plastid of the digested alga to profit from the photosynthesis; after a while, the plastids are also digested. This process is known as kleptoplasty, from the Greek, *kleptes*, thief.

Plant Hormone

Plant hormones (also known as phytohormones) are chemicals that regulate plant growth, which, in the UK, are termed 'plant growth substances'.

Plant hormones are signal molecules produced within the plant, and occur in extremely low concentrations. Hormones regulate cellular processes in targeted cells locally and, when moved to other locations, in other locations of the plant. Hormones also determine the formation of flowers, stems, leaves, the shedding of leaves, and the development and ripening of fruit. Plants, unlike animals, lack glands that produce and secrete hormones.

Instead, each cell is capable of producing hormones. Plant hormones shape the plant, affecting seed growth, time of flowering, the sex of flowers, senescence of leaves, and fruits. They affect which tissues grow upward and which grow downward, leaf formation and stem growth, fruit development and ripening, plant longevity, and even plant death. Hormones are vital to plant growth, and, lacking them, plants would be mostly a mass of undifferentiated cells. So they are also known as growth factors or growth hormones. The term 'Phytohormone' was coined by Thimann in 1948.

Phytohormones are found not only in higher plants, but in algae too.

Characteristics

The word hormone is derived from Greek, meaning *set in motion.* Plant hormones affect gene expression and transcription levels, cellular division, and growth. They are naturally produced within plants, though very similar chemicals are produced by fungi and bacteria that can also affect plant growth. A large number of related chemical compounds are synthesized by humans. They are used to regulate the growth of cultivated plants, weeds, and in vitro-grown plants and plant cells; these manmade compounds are called Plant Growth Regulators or PGRs for short. Early in the study of plant hormones, "phytohormone" was the commonly used term, but its use is less widely applied now.

Plant hormones are not nutrients, but chemicals that in small amounts promote and influence the growth, development, and differentiation of cells and tissues.

The biosynthesis of plant hormones within plant tissues is often diffuse and not always localized. Plants lack glands to produce and store hormones, because, unlike animals — which have two circulatory systems (lymphatic and cardiovascular) powered by a heart that moves fluids around the body — plants use more passive means to move chemicals around the plant. Plants utilize simple chemicals as hormones, which move more easily through the plant's tissues. They are often produced and used on a local basis within the plant body. Plant cells produce hormones that affect even different regions of the cell producing the hormone.

Hormones are transported within the plant by utilizing four types of movements. For localized movement, cytoplasmic streaming within cells and slow diffusion of ions and molecules between cells are utilized. Vascular tissues are used to move hormones from one part of the plant to another; these include sieve tubes or phloem that move sugars from the leaves to the roots and flowers, and xylem that moves water and mineral solutes from the roots to the foliage.

Not all plant cells respond to hormones, but those cells that do are programmed to respond at specific points in their growth cycle. The greatest effects occur at specific stages during the cell's life, with diminished effects occurring before or after this period. Plants need hormones at very specific times during plant growth and at specific locations. They also need to disengage the effects that hormones have when they are no longer needed. The production of hormones occurs very often at sites of active growth within the meristems, before cells have fully differentiated. After production, they are sometimes moved to other parts of the plant, where they cause an immediate effect; or they can be stored in cells to be released later. Plants use different pathways to regulate internal hormone quantities and moderate their effects; they can regulate the amount of chemicals used to biosynthesize hormones. They can store them in cells, inactivate them, or cannibalise already-formed hormones by conjugating them with carbohydrates, amino acids, or peptides. Plants can also break down hormones chemically, effectively destroying them. Plant hormones frequently regulate the concentrations of other plant hormones. Plants also move hormones around the plant diluting their concentrations.

The concentration of hormones required for plant responses are very low (10^{-6} to 10^{-5} mol/L). Because of these low concentrations, it has been very difficult to study plant hormones, and only since the late 1970s have scientists been able to start piecing together their effects and relationships to plant physiology. Much of the early work on plant hormones involved studying plants that were genetically deficient in one or involved the use of tissue-cultured plants grown *in vitro* that were subjected to differing ratios of hormones, and the resultant growth compared. The earliest scientific observation and study dates to the 1880s; the determination and observation of plant hormones and their identification was spread-out over the next 70 years.

Classes of Plant Hormones

In general, it is accepted that there are five major classes of plant hormones, some of which are made up of many different chemicals that can vary in structure from one plant to the next. The chemicals are each grouped together into one of these classes based on their structural similarities and on their effects on plant physiology. Other plant hormones and growth regulators are not easily grouped into these classes; they exist naturally or are synthesized by humans or other organisms, including chemicals that inhibit plant growth or interrupt the physiological processes within plants. Each class has positive as well as inhibitory functions, and most often work in tandem with each other, with varying ratios of one or more interplaying to affect growth regulation.

The five major classes are:

Abscisic Acid: Abscisic acid (also called ABA) is one of the most important plant growth regulators. It was discovered and researched under two different names before its chemical properties were fully known, it was called *dormin* and *abscicin II*. Once it was determined that the two compounds are the same, it was named abscisic acid. The name "abscisic acid" was given because it was found in high concentrations in newly abscissed or freshly fallen leaves.

This class of PGR is composed of one chemical compound normally produced in the leaves of plants, originating from chloroplasts, especially when plants are under stress. In general, it acts as an inhibitory chemical compound that affects bud growth, and seed and bud dormancy. It mediates changes within the apical meristem, causing bud dormancy and the alteration of the last set of leaves into protective bud covers. Since it was found in freshly abscissed leaves, it was thought to play a role in the processes of natural leaf drop, but further research has disproven this. In plant species from temperate parts of the world, it plays a role in leaf and seed dormancy by inhibiting growth, but, as it is dissipated from seeds or buds, growth begins. In other plants, as ABA levels decrease, growth then commences as gibberellin levels increase. Without ABA, buds and seeds would start to grow during warm periods in winter and be killed when it froze again. Since ABA dissipates slowly from the tissues and its effects take time to be offset by other plant hormones, there is a delay in physiological pathways that provide some protection from premature growth. It accumulates within seeds during fruit maturation, preventing seed germination within the fruit, or seed germination before winter. Abscisic acid's effects are degraded within plant tissues during cold temperatures or by its removal by water washing in out of the tissues, releasing the seeds and buds from dormancy.

In plants under water stress, ABA plays a role in closing the stomata. Soon after plants are water-stressed and the roots are deficient in water, a signal moves up to the leaves, causing the formation of ABA precursors there, which then move to the roots. The roots then release ABA, which is translocated to the foliage through the vascular system and modulates the potassium and sodium uptake within the guard cells, which then lose turgidity, closing the stomata. ABA exists in all parts of the plant and its concentration within any tissue seems to mediate its effects and function as a hormone; its degradation, or more properly catabolism, within the plant affects metabolic reactions and cellular growth and production of other hormones. Plants start life as a seed with high ABA levels. Just before the seed germinates, ABA levels decrease; during germination and early growth of the seedling, ABA levels decrease even more. As plants begin to produce shoots with fully functional leaves, ABA levels begin to increase, slowing down cellular growth in more "mature" areas of the plant. Stress from water or predation affects ABA production and catabolism rates, mediating another cascade of effects that trigger specific responses from targeted cells. Scientists are still piecing together the complex interactions and effects of this and other phytohormones.

Cytokinins: Cytokinins or CKs are a group of chemicals that influence cell division and shoot formation. They were called kinins in the past when the first cytokinins were isolated from yeast cells. They also help delay senescence or the aging of tissues, are responsible for mediating auxin transport throughout the plant, and affect internodal length and leaf growth. They have a highly synergistic effect in concert with auxins, and the ratios of these two groups of plant hormones affect most major growth periods during a plant's lifetime. Cytokinins counter the apical dominance induced by auxins; they in conjunction with ethylene promote abscission of leaves, flower parts, and fruits. The correlation of auxins and cytokinins in the plants is a constant (A/C = const.).

Ethylene: Ethylene is a gas that forms through the breakdown of methionine, which is in all cells. Ethylene has very limited solubility in water and does not accumulate within the cell but diffuses out of the cell and escapes out of the plant. Its effectiveness as a plant hormone is dependent on its rate of production versus its rate of escaping into the atmosphere. Ethylene is produced at a faster rate in rapidly growing and dividing cells, especially in darkness. New growth and newly germinated seedlings produce more ethylene than can escape the plant, which leads to elevated amounts of ethylene, inhibiting leaf expansion. As the new shoot is exposed to light, reactions by phytochrome in the plant's cells produce a signal for ethylene production to decrease, allowing leaf expansion. Ethylene affects cell growth and cell shape; when a growing shoot hits an obstacle while underground, ethylene production greatly increases, preventing cell elongation and causing the stem to swell. The resulting thicker stem can exert more pressure against the object impeding its path to the surface. If the shoot does not reach the surface and the ethylene stimulus becomes prolonged, it affects the stem's natural geotropic response, which is to grow upright, allowing it to grow around an object. Studies seem to indicate that ethylene affects stem diameter and height: When stems of trees are subjected to wind, causing lateral stress, greater ethylene production occurs, resulting in thicker, more sturdy tree trunks and branches. Ethylene affects fruit-ripening: Normally, when the seeds are mature, ethylene production increases and builds-up within the fruit, resulting in a climacteric event just before seed dispersal. The nuclear protein Ethylene Insensitive2 (EIN2) is regulated by ethylene production, and, in turn, regulates other hormones including ABA and stress hormones.

Gibberellins: Gibberellins, or GAs, include a large range of chemicals that are produced naturally within plants and by fungi. They were first discovered when Japanese researchers, including Eiichi Kurosawa, noticed a chemical produced by a fungus called *Gibberella fujikuroi* that produced abnormal growth in rice plants. Gibberellins are important in seed germination, affecting enzyme production that mobilizes food production used for growth of new cells. This is done by modulating chromosomal transcription. In grain (rice, wheat, corn, etc.) seeds, a layer of cells called the aleurone layer wraps around the endosperm tissue. Absoption of water by the seed causes production of GA. The GA is transported to the aleurone layer, which responds by producing enzymes that break down stored food reserves within the endosperm, which are utilized by the growing seedling. GAs produce bolting of rosette-forming plants, increasing internodal length. They promote flowering, cellular division, and in seeds growth after germination. Gibberellins also reverse the inhibition of shoot growth and dormancy induced by ABA.

Other Known Hormones: Other identified plant growth regulators include:

- Brassinosteroids, are a class of polyhydroxysteroids, a group of plant growth regulators. Brassinosteroids have been recognized as a sixth class of plant hormones, which stimulate cell elongation and division, gravitropism, resistance to stress, and xylem differentiation. They inhibit root growth and leaf abscission. Brassinolide was the first identified brassinosteroid and was isolated from extracts of rapeseed (*Brassica napus*) pollen in 1979.

- Salicylic acid — activates genes in some plants that produce chemicals that aid in the defence against pathogenic invaders.
- Jasmonates — are produced from fatty acids and seem to promote the production of defence proteins that are used to fend off invading organisms. They are believed to also have a role in seed germination, and affect the storage of protein in seeds, and seem to affect root growth.
- Plant peptide hormones — encompasses all small secreted peptides that are involved in cell-to-cell signalling. These small peptide hormones play crucial roles in plant growth and development, including defence mechanisms, the control of cell division and expansion, and pollen self-incompatibility.
- Polyamines — are strongly basic molecules with low molecular weight that have been found in all organisms studied thus far. They are essential for plant growth and development and affect the process of mitosis and meiosis.
- Nitric oxide (NO) — serves as signal in hormonal and defence responses (e.g. stomatal closure, root development, germination, nitrogen fixation, cell death, stress response). NO can be produced by a yet undefined NO synthase, a special type of nitrite reductase, nitrate reductase, mitochondrial cytochrome c oxidase or non enzymatic processes and regulate plant cell organelle functions (e.g. ATP synthesis in chloroplasts and mitochondria).
- Strigolactones, implicated in the inhibition of shoot branching.
- Karrikins, a group of plant growth regulators found in the smoke of burning plant material that have the ability to stimulate the germination of seeds.

Potential Medical Applications

Plant stress hormones activate cellular responses, including cell death, to diverse stress situations in plants. Researchers have found that some plant stress hormones share the ability to adversely affect human cancer cells. For example, sodium salicylate has been found to suppress proliferation of lymphoblastic leukemia, prostate, breast, and melanoma human cancer cells. Jasmonic acid, a plant stress hormone that belongs to the jasmonate family, induced death in lymphoblastic leukemia cells. Methyl jasmonate has been found to induce cell death in a number of cancer cell lines.

Auxin

Auxins (plural of auxin /ÈTÐk-sYn/) are a class of plant hormones (or plant growth substances) with some morphogen-like characteristics. Auxins have a cardinal role in coordination of many growth and behavioural processes in the plant's life cycle and are essential for plant body development. Auxins and their role in plant growth were first described by the Dutch scientist Frits Went. Kenneth V. Thimann isolated this phytohormone and determined its chemical structure as indole-3-acetic acid. Went and Thimann then co-authored a book on plant hormones, *Phytohormones*, in 1937.

Overview

The (dynamic and environment responsive) pattern of auxin distribution within the plant is a key factor for plant growth, its reaction to its environment, and specifically for development of plant organs (such as leaves or flowers). It is achieved through very complex and well coordinated active transport of auxin molecules from cell to cell throughout the plant body —by the so-called polar auxin transport. Thus, a plant can (as a whole) react to external conditions and adjust to them, without requiring a nervous system. Auxins typically act in concert with, or in opposition to, other plant hormones. For example, the ratio of auxin to cytokinin in certain plant tissues determines initiation of root versus shoot buds.

On the molecular level, all auxins are compounds with an aromatic ring and a carboxylic acid group. The most important member of the auxin family is indole-3-acetic acid (IAA). IAA generates the majority of auxin effects in intact plants, and is the most potent native auxin. And as native auxin, its stability is controlled in many ways in plants, from synthesis, through possible conjugation to degradation of its molecules, always according to the requirements of the situation. However, molecules of IAA are chemically labile in aqueous solution, so it is not used commercially as a plant growth regulator.

- The four naturally occurring (endogenous) auxins are IAA, 4-chloroindole-3-acetic acid, phenylacetic acid and indole-3-butyric acid; only these four were found to be synthesized by plants. However, most of the knowledge described so far in auxin biology and as described in the article below, apply basically to IAA; the other three endogenous auxins seems to have rather marginal importance for intact plants in natural environments. Alongside endogenous auxins, scientists and manufacturers have developed many synthetic compounds with auxinic activity.

- Synthetic auxin analogs include 1-naphthaleneacetic acid, 2,4-dichlorophenoxyacetic acid (2,4-D), and many others.

Some synthetic auxins, such as 2,4-D and 2,4,5-trichlorophenoxyacetic acid (2,4,5-T), are used also as herbicides. Broad-leaf plants (dicots), such as dandelions, are much more susceptible to auxins than narrow-leaf plants (monocots) such as grasses and cereal crops, so these synthetic auxins are valuable as synthetic herbicides.

Auxins are also often used to promote initiation of adventitious roots, and are the active ingredient of the commercial preparations used in horticulture to root stem cuttings. They can also be used to promote uniform flowering and fruit set, and to prevent premature fruit drop.

Discovery of Auxin

Charles Darwin: In 1881, Charles Darwin and his son Francis performed experiments on coleoptiles, the sheaths enclosing young leaves in germinating grass seedlings. Darwin and Francis experiment exposed the coleoptile to light from a unidirectional source and observed that they bend towards the light.

By covering various parts of the coleoptiles with a light impermeable opaque cap, the Darwin's discovered light is detected by coleoptile tip but that bending occurs in the hypocotyl. However the seedlings showed no signs of development towards light if the tip was covered with an opaque cap or if the tip was removed. The Darwin's concluded that tip of the coleoptile was responsible for sensing light. They proposed that a messenger is transmitted in a downward direction from the tip of the coleoptile, causing it to bend.

Peter Boysen-Jensen

In 1913 a Danish scientist named Peter Boysen-Jensen demonstrated that the signal was not transfixed but mobile. He separated the tip from the remainder of the coleoptile by a cube of gelatine which prevented cellular contact but, allowed chemicals to pass through. The seedlings responded normally bending towards the light. However when the tip was separated by an impermeable substance, there was no curvature of the stem.

Friedrich Went

In 1926, Friedrich Went, showed that a chemical messenger diffuses from coleoptiles tips. Went's experiment identified how a growth promoting chemical causes a coleoptile to grow towards light. Went cut the tips of the coleoptiles and placed them in the dark, putting a few tips on agar blocks that he predicted would absorb the growth-promoting chemical. On control coleoptiles, he placed a block that lacked the chemical. On others, he placed blocks containing the chemical, either centred on top of the coleoptile to distribute the chemical evenly or offset to increase the concentration on one side. When the growth promoting chemical was distributed evenly the coleoptile grew straight. If the chemical was distributed unevenly, the coleoptile curved away from the side with the cube, as if growing towards light, even though it was grown in the dark. Went later proposed that the messenger substance is a growth-promoting hormone, which he named auxin, that becomes asymmetrically distributed in the bending region. Went concluded that auxin is at a higher concentration on the shaded side, promoting cell elongation, which results in a coleoptiles bending towards the light.

Hormonal Activity

Auxins coordinate development at all levels in plants, from the cellular level, through organs, and ultimately to the whole plant.

Molecular Mechanisms

Auxin molecules present in cells may trigger responses directly through stimulation or inhibition of the expression of sets of certain genes or by means independent of gene expression. Auxin transcriptionally activates four different families of *early genes* (aka *primary response genes*), so-called because the components required for the activation are preexisting, leading to a rapid response. The families are glutathione S-transferases, auxin homeostasis proteins like GH3, SAUR genes of currently unknown function, and the Aux/IAA repressors.

Aux/IAA, ARF, TIR1, SCF Auxin Regulatory Pathways

The Aux/IAA repressors provide an example of one of the pathways leading to auxin induced changes of gene expression. This pathway involves the protein families TIR1 (transport inhibitor response1),

ARF (auxin response factor), Aux/IAA transcriptional repressors, and the ubiquitin ligase complex that is a part of the ubiquitin-proteasome protein degradation pathway. ARF proteins have DNA binding domains and can bind promoter regions of genes and activate or repress gene expression. Aux/IAA proteins can bind ARF proteins sitting on gene promoters and prevent them from doing their job. TIR1 proteins are F-box proteins that have three different domains giving them the ability to bind to three different ligands: an SCF ubiquitin ligase complex (using the F-box domain), auxin (so TIR1 proteins are auxin receptors), and Aux/IAA proteins (via a degron domain). Upon binding of auxin, a TIR1 protein's degron domain has increased affinity for Aux/IAA repressor proteins, which when bound to TIR1 and its SCF complex undergo ubiquitination and subsequent degradation by a proteasome. The degradation of Aux/IAA proteins frees ARF proteins to activate or repress genes at whose promoters they are bound.

Within a plant, elaboration of the Aux/IAA repressor pathway takes place via diversification of the TIR1, ARF, and Aux/IAA protein families. Each family may contain many similar-acting proteins, differing in qualities such as degree of affinity for partner proteins, amount of activation or repression of target gene transcription, or domains of expression (e.g. different plant tissues might express different members of the family, or different environmental stresses might activate expression of different members). Such elaboration permits the plant to use auxin in a variety of ways depending on the needs of the tissue and plant.

Other Auxin Regulatory Pathways

Another protein, auxin-binding protein 1 (ABP1), is a putative receptor for a different signalling pathway, but its role is as yet unclear. Electrophysiological experiments with protoplasts and anti-ABP1 antibodies suggest ABP1 may have a function at the plasma membrane, and cells can possibly use ABP1 proteins to respond to auxin through means faster and independent of gene expression.

On a Cellular Level

On the cellular level, auxin is essential for cell growth, affecting both cell division and cellular expansion. Auxin concentration level, together with other local factors, contributes to cell differentiation and specification of the cell fate.

Depending on the specific tissue, auxin may promote axial elongation (as in shoots), lateral expansion (as in root swelling), or isodiametric expansion (as in fruit growth). In some cases (coleoptile growth), auxin-promoted cellular expansion occurs in the absence of cell division. In other cases, auxin-promoted cell division and cell expansion may be closely sequenced within the same tissue (root initiation, fruit growth). In a living plant, auxins and other plant hormones nearly always appear to interact to determine patterns of plant development.

Organ Patterns

Growth and division of plant cells together result in growth of tissue, and specific tissue growth contributes to the development of plant organs.

Growth of cells contributes to the plant's size, unevenly localized growth produces bending, turning and directionalization of organs-for example, stems turning toward light sources (phototropism), roots growing in response to gravity (gravitropism), and other tropisms originated because cells on one side grow faster than the cells on the other side of the organ. So, precise control of auxin distribution between different cells has paramount importance to the resulting form of plant growth and organization.

Auxin Transport and the Uneven Distribution of Auxin

To cause growth in the required domains, auxins must of necessity be active preferentially in them. Auxins are not synthesized in all cells (even if cells retain the potential ability to do so, only under specific conditions will auxin synthesis be activated in them). For that purpose, auxins have to be not only translocated toward those sites where they are needed, but also they must have an established mechanism to detect those sites. For that purpose, auxins have to be translocated toward those sites where they are needed. Translocation is driven throughout the plant body, primarily from peaks of shoots to peaks of roots (from up to down).

For long distances, relocation occurs via the stream of fluid in phloem vessels, but, for short-distance transport, a unique system of coordinated polar transport directly from cell to cell is exploited. This short-distance, active transport exhibits some morphogenetic properties.

This process, the polar auxin transport, is directional, very strictly regulated, and based in uneven distribution of auxin efflux carriers on the plasma membrane, which send auxins in the proper direction. Pin-formed (PIN) proteins are vital in transporting auxin.

The regulation of PIN protein localisation in a cell determines the direction of auxin transport from cell, and concentrated effort of many cells creates peaks of auxin, or auxin maxima (regions having cells with higher auxin - a maximum). Proper and timely auxin maxima within developing roots and shoots are necessary to organise the development of the organ. Surrounding auxin maxima are cells with low auxin troughs, or auxin minima. For example, in the *Arabidopsis* fruit, auxin minima have been shown to be important for its tissue development.

Organization of the Plant

As auxins contribute to organ shaping, they are also fundamentally required for proper development of the plant itself. Without hormonal regulation and organization, plants would be merely proliferating heaps of similar cells. Auxin employment begins in the embryo of the plant, where directional distribution of auxin ushers in subsequent growth and development of primary growth poles, then forms buds of future organs. Next, it helps to coordinate proper development of the arising organs, such as roots, cotyledons and leaves and mediates long distance signals between them, contributing so to the overall architecture of the plant. Throughout the plant's life, auxin helps the plant maintain the polarity of growth, and actually "recognize" where it has its branches (or any organ) connected.

An important principle of plant organization based upon auxin distribution is apical dominance, which means the auxin produced by the apical bud (or growing tip) diffuses (and is transported) downwards and inhibits the development of ulterior lateral bud growth, which would otherwise compete with the apical tip for light and nutrients. Removing the apical tip and its suppressively acting auxin allows the lower dormant lateral buds to develop, and the buds between the leaf stalk and stem produce new shoots which compete to become the lead growth. The process is actually quite complex, because auxin transported downwards from the lead shoot tip has to interact with several other plant hormones (such as strigolactones or cytokinins) in the process on various positions along the growth axis in plant body to achieve this phenomenon. This plant behaviour is used in pruning by horticulturists.

Finally, the sum of auxin arriving from stems to roots influences the degree of root growth. If shoot tips are removed, the plant does not react just by outgrowth of lateral buds — which are supposed to replace to original lead. It also follows that smaller amount of auxin arriving to the roots results in slower growth of roots and the nutrients are subsequently in higher degree invested in the upper part of the plant, which hence starts to grow faster.

The Five Models for Auxin Transport

There have been five models made for auxin transport in the plant Arabidopsis.

First Model

In the first model incoming light deactivates auxin on the light side of the plant allowing the shaded part to continue growing and eventually bend the plant over towards the light.

Second Model

In the second model light inhibits auxin biosynthesis on the light side of the plant, thus decreasing the concentration of auxin relative to the unaffected side.

Third Model

In the third model there is a horizontal flow of auxin from both the light and dark side of the plant. Incoming light causes more auxin to flow from the exposed side to the shaded side, increasing the concentration of auxin on the shaded side and thus growth occurring.

Fourth Model

In the fourth model shows receiving light to inhibit auxin basipetal down to the exposed side, causing the auxin to only flow down the shaded side.

Fifth Model

Model five encompasses elements of both model 3 and 4. The main auxin flow in this model comes from the top of the plant vertically down towards the base of the plant with some of the auxin travelling horizontally from the main auxin flow to both sides of the plant. Receiving light inhibits the horizontal auxin flow from the main vertical auxin flow to the irradiated exposed side.

Effects

Auxin participates in phototropism, geotropism, hydrotropism and other developmental changes. The uneven distribution of auxin, due to environmental cues, such as unidirectional light or gravity force, results in uneven plant tissue growth, and generally, auxin governs the form and shape of plant body, direction and strength of growth of all organs, and their mutual interaction.

Auxin stimulates cell elongation by stimulating wall-loosening factors, such as elastins, to loosen cell walls. The effect is stronger if gibberellins are also present. Auxin also stimulates cell division if cytokinins are present. When auxin and cytokinin are applied to callus, rooting can be generated if the auxin concentration is higher than cytokinin concentration. Xylem tissues can be generated when the auxin concentration is equal to the cytokinins.

Auxin also induces sugar and mineral accumulation at the site of application.

Wound Response

Auxin induces the formation and organization of phloem and xylem. When the plant is wounded, the auxin may induce the cell differentiation and regeneration of the vascular tissues.

Root Growth and Development

Auxins promote root initiation. Auxin induces both growth of pre-existing roots and adventitious root formation, i.e., branching of the roots. As more native auxin is transported down the stem to the roots, the overall development of the roots is stimulated. If the source of auxin is removed, for example the tips of stems are trimmed, the roots are less stimulated accordingly, and growth of stem is supported instead.

In horticulture, auxins, especially NAA and IBA, are commonly applied to stimulate root initiation when rooting cuttings of plants. However, high concentrations of auxin inhibit root elongation and instead enhance adventitious root formation. Removal of the root tip can lead to inhibition of secondary root formation.

Apical Dominance

Auxin induces shoot apical dominance; the axillary buds are inhibited by auxin, as a high concentration of auxin directly stimulates ethylene synthesis in lateral buds, causing inhibition of their growth and potentiation of apical dominance. When the apex of the plant is removed, the inhibitory effect is removed and the growth of lateral buds is enhanced. Auxin is sent to the part of the plant facing light, and this promotes growth towards that direction.

Fruit Growth and Development

Auxin is required for fruit growth and development and delays fruit senescence. When seeds are removed from strawberries, fruit growth is stopped; exogenous auxin stimulates the growth in fruits with seeds removed. For fruit with unfertilized seeds, exogenous auxin results in parthenocarpy ("virgin-fruit" growth).

Fruits form abnormal morphologies when auxin transport is disturbed. In *Arabidopsis* fruits, auxin controls the release of seeds from the fruit (pod). The valve margins are a specialised tissue in pods that regulates when pod will open (dehiscence). Auxin must be removed from the valve margin cells to allow the valve margins to form. This process requires modification of the auxin transporters (PIN proteins).

Flowering

Auxin plays also a minor role in the initiation of flowering and development of reproductive organs. In low concentrations, it can delay the senescence of flowers.

Ethylene Biosynthesis

In low concentrations, auxin can inhibit ethylene formation and transport of precursor in plants; however, high concentrations can induce the synthesis of ethylene. Therefore, the high concentration can induce femaleness of flowers in some species.

Auxin inhibits abscission prior to formation of abscission layer, and thus inhibits senescence of leaves.

Synthetic Auxins

In the course of research on auxin biology, many compounds with noticeable auxin activity were synthesized. Many of them had been found to have economical potential for man-controlled growth and development of plants in agronomy. Synthetic auxins include the following compounds:

Auxins are toxic to plants in large concentrations; they are most toxic to dicots and less so to monocots. Because of this property, synthetic auxin herbicides, including 2,4-D and 2,4,5-T, have been developed and used for weed control.

However, synthetic auxins, especially 1-naphthaleneacetic acid (NAA) and indole-3-butyric acid (IBA), are also commonly applied to stimulate root growth when taking cuttings of plants or for different agricultural purposes such as the prevention of fruit drop in orchards.

Used in high doses, auxin stimulates the production of ethylene. Excess ethylene (also native plant hormone) can inhibit elongation growth, cause leaves to fall (abscission), and even kill the plant. Some synthetic auxins, such as 2,4-D and 2,4,5-trichlorophenoxyacetic acid (2,4,5-T) were marketed also as herbicides. Dicots, such as dandelions, are much more susceptible to auxins than monocots), such as grasses and cereal crops. So, these synthetic auxins are valuable as synthetic herbicides. 2,4-D was the first widely used herbicide, and it is still so. 2,4-D was first commercialized by the Sherwin-Williams company, and saw use in the late 1940s. It is easy and inexpensive to manufacture.

Herbicide Manufacture

The defoliant Agent Orange, used extensively by American forces in the Vietnam War, was a mix of 2,4-D and 2,4,5-T. The compound 2,4-D is still in use and is thought to be safe, but 2,4,5-T was more or less banned by the EPA in 1979. The dioxin TCDD is an unavoidable contaminant produced in the manufacture of 2,4,5-T. As a result of the integral dioxin contamination, 2,4,5-T has been implicated in leukemia, miscarriages, birth defects, liver damage, and other diseases.

Hormones and Plant Propagation

Synthetic plant hormones or PGRs are commonly used in a number of different techniques involving plant propagation from cuttings, grafting, micropropagation, and tissue culture.

The propagation of plants by cuttings of fully developed leaves, stems, or roots is performed by gardeners utilizing auxin as a rooting compound applied to the cut surface; the auxins are taken into the plant and promote root initiation. In grafting, auxin promotes callus tissue formation, which joins the surfaces of the graft together. In micropropagation, different PGRs are used to promote multiplication and then rooting of new plantlets. In the tissue-culturing of plant cells, PGRs are used to produce callus growth, multiplication, and rooting.

Seed Dormancy

A dormant seed is one that is unable to germinate in a specified period of time under a combination of environmental factors that are normally suitable for the germination of the non-dormant seed. Dormancy is a mechanism to prevent germination during unsuitable ecological conditions, when the probability of seedling survival is low.

One important function of most seeds is delayed germination, which allows time for dispersal and prevents germination of all the seeds at the same time. The staggering of germination safeguards some seeds and seedlings from suffering damage or death from short periods of bad weather or from transient herbivores; it also allows some seeds to germinate when competition from other plants for light and water might be less intense. Another form of delayed seed germination is seed quiescence, which is different than true seed dormancy and occurs when a seed fails to germinate because the external environmental conditions are too dry or warm or cold for germination. Many species of plants have seeds that delay germination for many months or years, and some seeds can remain in the soil seed bank for more than 50 years before germination. Some seeds have a very long viability period, and the oldest documented germinating seed was nearly 2000 years old based on radiocarbon dating.

Overview

True dormancy or innate dormancy is caused by conditions within the seed that prevent germination under normally ideal conditions. Often seed dormancy is divided into two major categories based on what part of the seed produces dormancy: exogenous and endogenous. There are three types of dormancy based on their mode of action: physical, physiological and morphological.

There have been a number of classification schemes developed to group different dormant seeds, but none have gained universal usage. Dormancy occurs because of a wide range of reasons that often overlap, producing conditions in which definitive categorization is not clear. Compounding this problem is that the same seed that is dormant for one reason at a given point may be dormant because of another reason at a later point. Some seeds fluctuate from periods of dormancy to non dormancy, and despite the fact that a dormant seed appears to be static or inert, in reality they are still receiving and responding to environmental cues.

Exogenous Dormancy

Exogenous dormancy is caused by conditions outside the embryo and is often broken down into three subgroups:

Physical Dormancy

Dormancy that is caused by an impermeable seed coat is known as physical dormancy. Physical dormancy is the result of impermeable layer(s) that develops during maturation and drying of the seed or fruit. This impermeable layer prevents the seed from taking up water or gases. As a result, the seed is prevented from germinating until dormancy is broken. In natural systems, physical dormancy is broken by several factors including high temperatures, fluctuating temperatures, fire, freezing/thawing, drying or passage through the digestive tracts of animals. Physical dormancy is believed to have developed >100 mya.

Except for a few cases, once physical dormancy is broken it cannot be reinstated i.e. the seed is unable to enter secondary dormancy following unfavourable conditions unlike seeds with physiological dormancy mechanisms. Therefore, the timing of the mechanisms that breaks physical dormancy is critical and must be tuned to environmental cues. This maximises the chances for germination occurring in conditions where the plant will successfully germinate, establish and eventually reproduce.

Physical dormancy has been identified in the seeds of plants across 15 angiosperm families including:

- Anacardiaceae
- Bixaceae
- Cannaceae (monocot)
- Cistaceae
- Cochlospermaceae
- Convolvulaceae
- Cucurbitaceae
- Dipterocarpaceae
- Geraniaceae
- Legumeinosae
- Malvaceae
- Nelumbonaceae
- Rhamnaceae
- Sarcolaenaceae
- Sapindaceae

Physical Dormancy Has Not been Recorded in Gymnosperms.

Generally, physical dormancy is the result of one or more palisade layers in the fruit or seed coat. These layers are lignified with malphigian cells tightly packed together and impregnated with water-repellent. In the Anacardiaceae and Nelumbonaceae families the seed coat is not well developed. Therefore, palisade layers in the fruit perform the functional role of preventing water uptake . While physical dormancy is a common feature, several species in these families do not have physical dormancy or produce non-dormant seeds.

Specialised structures, which function as a "water-gap", are associated with the impermeable layers of the seed to prevent the uptake of water. The water-gap is closed at seed maturity and is opened in response to the appropriate environmental signal. Breaking physical dormancy involves the disruption of these specialised structures within the seed, and acts as an environmental signal detector for germination. For example, legume (Fabaceae) seeds become permeable after the thin-walled cells of lens (water-gap structure}. Following disrupted pulls apart to allow water entry into the seed.Other water gap structures include carpellary micropyle, bixoid chalazal plug, imbibition lid and the suberised 'stopper'.

In nature, the seed coats of physically dormant seeds are thought to become water permeable over time through repeated heating and cooling over many months-years in the soil seedbank. For example, the high and fluctuating temperatures during the dry season in northern Australia promote dormancy break in impermeable seeds of *Stylosanthes humilis* and *S.hamata* (Fabaceae).

Mechanical Dormancy

Mechanical dormancy occurs when seed coats or other coverings are too hard to allow the embryo to expand during germination. In the past this mechanism of dormancy was ascribed to a number of species that have been found to have endogenous factors for their dormancy instead. These endogenous facts include physiologically dormancy cased by low embryo growth potential

Chemical Dormancy

Includes growth regulators etc., that are present in the coverings around the embryo. They may be leached out of the tissues by washing or soaking the seed, or deactivated by other means. Other chemicals that prevent germination are washed out of the seeds by rainwater or snow melt.

Endogenous Dormancy

Endogenous dormancy is caused by conditions within the embryo itself, and it is also often broken down into three subgroups: physiological dormancy, morphological dormancy and combined dormancy, each of these groups may also have subgroups.

Physiological Dormancy

Physiological dormancy prevents embryo growth and seed germination until chemical changes occur. These chemicals include inhibitors that often retard embryo growth to the point where it is not strong enough to break through the seed coat or other tissues. Physiological dormancy is indicated when an increase in germination rate occurs after an application of gibberellic acid (GA3) or after Dry after-ripening or dry storage. It is also indicated when dormant seed embryos are excised and produce healthy seedlings: or when up to 3 months of cold (0–10°C) or warm (=15°C) stratification increases germination: or when dry after-ripening shortens the cold stratification period required. In some seeds physiological dormancy is indicated when scarification increases germination.

Physiological dormancy is broken when inhibiting chemicals are broken down or are no longer produced by the seed; often by a period of cool moist conditions, normally below (+4C) 39F, or in the case of many species in *Ranunculaceae* and a few others,("5C) 24F. Abscisic acid is usually the growth inhibitor in seeds and its production can be affected by light.

Some plants like Peony species have multiple types of physiological dormancy, one affects radicle (root) growth while the other affects plumule (shoot) growth. Seeds with physiological dormancy most often do not germinate even after the seed coat or other structures that interfere with embryo growth are removed. Conditions that affect physiological dormancy of seeds include:

- Drying; some plants including a number of grasses and those from seasonally arid regions need a period of drying before they will germinate, the seeds are released but need to have a lower moisture content before germination can begin. If the seeds remain moist after dispersal, germination can be delayed for many months or even years. Many herbaceous plants from temperate climate zones have physiological dormancy that disappears with drying of the seeds. Other species will germinate after dispersal only under very narrow temperature ranges, but as the seeds dry they are able to germinate over a wider temperature range.
- Photodormancy or light sensitivity affects germination of some seeds. These photoblastic seeds need a period of darkness or light to germinate. In species with thin seed coats, light may be able to penetrate into the dormant embryo. The presence of light or the absence of light may trigger the germination process, inhibiting germination in some seeds buried too deeply or in others not buried in the soil.
- Thermodormancy is seed sensitivity to heat or cold. Some seeds including cocklebur and amaranth germinate only at high temperatures (30C or 86F) many plants that have seed

that germinate in early to mid summer have thermodormancy and germinate only when the soil temperature is warm. Other seeds need cool soils to germinate, while others like celery are inhibited when soil temperatures are too warm. Often thermodormancy requirements disappear as the seed ages or dries.

Seeds are classified as having deep physiological dormancy under these conditions: applications of GA3 does not increase germination; or when excised embryos produce abnormal seedlings; or when seeds require more than 3 months of cold stratification to germinate.

Morphological Dormancy

Embryo underdeveloped or undifferentiated. Some seeds have fully differentiated embryos that need to grow more before seed germination, or the embryos are not differentiated into different tissues at the time of fruit ripening.

- Immature embryos – some plants release their seeds before the tissues of the embryos have fully differentiated, and the seeds ripen after they take in water while on the ground, germination can be delayed from a few weeks to a few months.

Combined Dormancy

Seeds have both morphological and physiological dormancy.

- Morpho-physiological or morphophysiological dormancy occurs when seeds with underdeveloped embryos, also have physiological components to dormancy. These seeds therefore require dormancy-breaking treatments as well as a period of time to develop fully grown embryos.
- Intermediate simple
- Deep simple
- Deep simple epicotyl
- Deep simple double
- Intermediate complex
- Deep complex

Combinational Dormancy

Combinational dormancy occurs in some seeds, where dormancy is caused by both exogenous (physical) and endogenous (physiological) conditions. some *Iris* species have both hard impermeable seeds coats and physiological dormancy.

Secondary Dormancy

Secondary dormancy occurs in some non-dormant and post dormant seeds that are exposed to conditions that are not favourable for germination, like high temperatures. It is caused by conditions that occur after the seed has been dispersed. The mechanisms of secondary dormancy are not yet fully understood but might involve the loss of sensitivity in receptors in the plasma membrane.

Not all seeds undergo a period of dormancy, many species of plants release their seeds late in the year when the soil temperature is too low for germination or when the environment is dry. If these seeds are collected and sown in an environment that is warm enough, and/or moist enough, they will germinate. Under natural conditions non dormant seeds released late in the growing season wait until spring when the soil temperature rises or in the case of seeds dispersed during dry periods until it rains and there is enough soil moisture.

Seeds that do not germinate because they have fleshy fruits that retard germination are quiescent, not dormant.

Many garden plants have seeds that will germinate readily as soon as they have water and are warm enough, though their wild ancestors had dormancy. These cultivated plants lack seed dormancy because of generations of selective pressure by plant breeders and gardeners that grew and kept plants that lacked dormancy.

Seeds of some mangroves are viviparous and begin to germinate while still attached to the parent; they produce a large, heavy root, which allows the seed to penetrate into the ground when it falls.

Golgi Apparatus

The Golgi apparatus, also known as the Golgi complex, Golgi body, or simply the Golgi, is an organelle found in most eukaryotic cells. It was identified in 1897 by the Italian physician Camillo Golgi and named after him in 1898.

Part of the cellular endomembrane system, the Golgi apparatus packages proteins inside the cell before they are sent to their destination; it is particularly important in the processing of proteins for secretion.

Discovery

Owing to its large size, the Golgi apparatus was one of the first organelles to be discovered and observed in detail. It was discovered in 1898 by Italian physician Camillo Golgi during an investigation of the nervous system. After first observing it under his microscope, he termed the structure the *internal reticular apparatus*. Some doubted the discovery at first, arguing that the appearance of the structure was merely an optical illusion created by the observation technique used by Golgi.

With the development of modern microscopes in the 20th century, the discovery was confirmed. Early references to the Golgi referred to it by various names including the "Golgi–Holmgren apparatus", "Golgi–Holmgren ducts", and "Golgi–Kopsch apparatus". The term "Golgi apparatus" was used in 1910 and first appeared in scientific literature in 1913.

Structure

Found within the cytoplasm of both plant and animal cells, the Golgi is composed of stacks of membrane-bound structures known as cisternae (singular: *cisterna*). An individual stack is sometimes called a dictyosome, especially in plant cells. A mammalian cell typically contains 40 to 100 stacks. Between four and eight cisternae are usually present in a stack; however, in some protists as many as sixty have been observed. Each cisterna comprises a flat, membrane enclosed disc that includes special Golgi enzymes which modify or help to modify cargo proteins that travel through it.

The cisternae stack has four functional regions: the cis-Golgi network, medial-Golgi, endo-Golgi, and trans-Golgi network. Vesicles from the endoplasmic reticulum (via the vesicular-tubular clusters) fuse with the network and subsequently progress through the stack to the trans Golgi network, where they are packaged and sent to their destination. Each region contains different enzymes which selectively modify the contents depending on where they reside. The cisternae also carry structural proteins important for their maintenance as flattened membranes which stack upon each other.

Function of a Golgi Body

Cells synthesize a large number of different macromolecules. The Golgi apparatus is integral in modifying, sorting, and packaging these macromolecules for cell secretion (exocytosis) or use within the cell. It primarily modifies proteins delivered from the rough endoplasmic reticulum but is also involved in the transport of lipids around the cell, and the creation of lysosomes. In this respect it can be thought of as similar to a post office; it packages and labels items which it then sends to different parts of the cell.

Enzymes within the cisternae are able to modify the proteins by addition of carbohydrates (glycosylation) and phosphates (phosphorylation). In order to do so, the Golgi imports substances such as nucleotide sugars from the cytosol. These modifications may also form a signal sequence which determines the final destination of the protein. For example, the Golgi apparatus adds a mannose-6-phosphate label to proteins destined for lysosomes.

The Golgi plays an important role in the synthesis of proteoglycans, which are molecules present in the extracellular matrix of animals. It is also a major site of carbohydrate synthesis. This includes the production of glycosaminoglycans (GAGs), long unbranched polysaccharides which the Golgi then attaches to a protein synthesised in the endoplasmic reticulum to form proteoglycans. Enzymes in the Golgi polymerize several of these GAGs via a xylose link onto the core protein. Another task of the Golgi involves the sulfation of certain molecules passing through its lumen via sulfotranferases that gain their sulphur molecule from a donor called PAPS. This process occurs on the GAGs of proteoglycans as well as on the core protein. Sulfation is generally performed in the trans-Golgi network. The level of sulfation is very important to the proteoglycans' signalling abilities as well as giving the proteoglycan its overall negative charge.

The phosphorylation of molecules requires that ATP is imported into the lumen of the Golgi and utilised by resident kinases such as casein kinase 1 and casein kinase 2. One molecule that is phosphorylated in the Golgi is Apolipoprotein, which forms a molecule known as VLDL that is a constituent of blood serum. It is thought that the phosphorylation of these molecules is important to help aid in their sorting for secretion into the blood serum.

The Golgi has a putative role in apoptosis, with several Bcl-2 family members localised there, as well as to the mitochondria. A newly characterized protein, GAAP (Golgi anti-apoptotic protein), almost exclusively resides in the Golgi and protects cells from apoptosis by an as-yet undefined mechanism.

Vesicular Transport

The vesicles that leave the rough endoplasmic reticulum are transported to the *cis* face of the Golgi apparatus, where they fuse with the Golgi membrane and empty their contents into the lumen. Once inside the lumen, the molecules are modified, then sorted for transport to their next destinations. The Golgi apparatus tends to be larger and more numerous in cells that synthesize and secrete large amounts of substances; for example, the plasma B cells and the antibody-secreting cells of the immune system have prominent Golgi complexes.

Those proteins destined for areas of the cell other than either the endoplasmic reticulum or Golgi apparatus are moved towards the *trans* face, to a complex network of membranes and associated vesicles known as the *trans-Golgi network* (TGN). This area of the Golgi is the point at which proteins are sorted and shipped to their intended destinations by their placement into one of at least three different types of vesicles, depending upon the molecular marker they carry.

Types	*Description*	*Example*
Exocytotic vesicles *(continuous)*	Vesicle contains proteins destined for extracellular release. After packaging, the vesicles bud off and immediately move towards the plasma membrane, where they fuse and release the contents into the extracellular space in a process known as *constitutive secretion*.	Antibody release by activated plasma B cells
Secretory vesicles *(regulated)*	Vesicle contains proteins destined for extracellular release. After packaging, the vesicles bud off and are stored in the cell until a signal is given for their release. When the appropriate signal is received they move towards the membrane and fuse to release their contents. This process is known as *regulated secretion*.	Neurotransmitter release from neurons
Lysosomal vesicles	Vesicle contains proteins and ribosomes destined for the lysosome, an organelle of degradation containing many acid hydrolases, or to lysosome-like storage organelles. These proteins include both digestive enzymes and membrane proteins. The vesicle first fuses with the late endosome, and the contents are then transferred to the lysosome via unknown mechanisms.	Digestive proteases destined for the lysosome

Transport Mechanism

The transport mechanism which proteins use to progress through the Golgi apparatus is not yet clear; however a number of hypotheses currently exist. Until recently, the vesicular transport mechanism was favoured but now more evidence is coming to light to support cisternal maturation. The two proposed models may actually work in conjunction with each other, rather than being mutually exclusive. This is sometimes referred to as the *combined* model.

- *Cisternal maturation model*: the cisternae of the Golgi apparatus move by being built at the *cis* face and destroyed at the *trans* face. Vesicles from the endoplasmic reticulum fuse with each other to form a cisterna at the *cis* face, consequently this cisterna would appear to move through the Golgi stack when a new cisterna is formed at the *cis* face. This model is supported by the fact that structures larger than the transport vesicles, such as collagen rods, were observed microscopically to progress through the Golgi apparatus. This was initially a popular hypothesis, but lost favour in the 1980s. Recently it has made a comeback, as laboratories at the University of Chicago and the University of Tokyo have been able to use new technology to directly observe Golgi compartments maturing. Additional evidence comes from the fact that COPI vesicles move in the retrograde direction, transporting endoplasmic reticulum proteins back to where they belong by recognizing a signal peptide.

- *Vesicular transport model*: Vesicular transport views the Golgi as a very stable organelle, divided into compartments in the cis to trans direction. Membrane bound carriers transport material between the endoplasmic reticulum and the different compartments of the Golgi. Experimental evidence includes the abundance of small vesicles (known technically as shuttle vesicles) in proximity to the Golgi apparatus. To direct the vesicles, actin filaments connect packaging proteins to the membrane to ensure that they fuse with the correct compartment.

Fate During Mitosis

In animal cells, the Golgi apparatus will break up and disappear following the onset of mitosis, or cellular division. During the telophase of mitosis, the Golgi apparatus reappears. As of December 2009 it is uncertain how this occurs. In contrast, Golgi stacks have been observed to remain intact in plant or yeast cells throughout the cell cycle. The reason for this difference is not yet known, but it may, in part, be a consequence of golgin proteins.

Microfilament

Microfilaments or actin filaments are the thinnest filaments of the cytoskeleton, a structure found in the cytoplasm of eukaryotic cells. These linear polymers of actin subunits are flexible and relatively strong, resisting buckling by multi-piconewton compressive forces and filament fracture by nanonewton tensile forces. Microfilaments are highly versatile, functioning in cytokinesis, amoeboid movement, and changes in cell shape. In inducing this cell motility, one end of the actin filament elongates while the other end contracts, presumably by myosin II molecular motors. Additionally, they function as part of actomyosin-driven contractile molecular motors, wherein the thin filaments serve as tensile platforms for myosin's ATP-dependent pulling action in muscle contraction and pseudopod advancement. Microfilaments have a tough, flexible framework which helps the cell in movement.

Organization

Actin filaments are assembled in two general types of structures: bundles and networks. Bundles can be composed of polar filament arrays, in which all barbed ends point to the same end of the bundle, or non-polar arrays, where the barbed ends point towards both ends. A class of actin-binding proteins, called cross-linking proteins, dictate the formation of these structures. Cross-linking proteins determine filament orientation and spacing in the bundles and networks. These structures are regulated by many other classes of actin-binding proteins, including motor proteins, branching proteins, severing proteins, polymerization promoters, and capping proteins.

In Vitro *Self-assembly*

Measuring approximately 6 nm in diameter, microfilaments are the thinnest fibres of the cytoskeleton. They are polymers of actin subunits (globular actin, or G-actin), which as part of the fibre are referred to as filamentous actin, or F-actin. Each microfilament is made up of two helical, interlaced strands of subunits. Much like microtubules, actin filaments are polarized. Electron micrographs have provided evidence of their fast-growing barbed-ends and their slow-growing pointed-end. This polarity has been determined by the pattern created by the binding of myosin S1 fragments: they themselves subunits of the larger myosin II protein complex. The pointed end is commonly referred to as the minus (–) end and the barbed end is referred to as the plus (+) end.

In vitro actin polymerization, or nucleation, starts with the self-association of three G-actin monomers to form a trimer. ATP-bound actin then itself binds the barbed end, and the ATP is subsequently hydrolyzed. ATP hydrolysis occurs with a half time of about 2 seconds, while the half time for the dissociation of the inorganic phosphate is about 6 minutes. This autocatalyzed event reduces the binding strength between neighbouring subunits, and thus generally destabilizes the filament. *In vivo* actin polymerization is catalyzed by a class of filament end-tracking molecular motors known as actoclampins. Recent evidence suggests that the rate of ATP hydrolysis and the rate of monomer incorporation are strongly coupled.

Subsequently, ADP-actin dissociates slowly from the pointed end, a process significantly accelerated by the actin-binding protein, cofilin. ADP bound cofilin severs ADP-rich regions nearest the (–)-ends. Upon release, the free actin monomer slowly dissociates from ADP, which in turn rapidly binds to the free ATP diffusing in the cytosol, thereby forming the ATP-actin monomeric units needed for further barbed-end filament elongation. This rapid turnover is important for the cell's movement. End-capping proteins such as CapZ prevent the addition or loss of monomers at the filament end where actin turnover is unfavourable, such as in the muscle apparatus.

Actin polymerization together with capping proteins were recently used to control the 3-dimensional growth of protein filament so as to perform 3D topologies useful in technology and the making of electrical interconnect. Electrical conductivity is obtained by metallisation of the protein 3D structure.

Mechanism of Force Generation

As a result of ATP hydrolysis, filaments elongate approximately 10 times faster at their barbed ends than their pointed ends. At steady-state, the polymerization rate at the barbed end matches the depolymerization rate at the pointed end, and microfilaments are said to be treadmilling. Treadmilling results in elongation in the barbed end and shortening in the pointed-end, so that the filament in total moves. Since both processes are energetically favourable, this means force is generated, the energy ultimately coming from ATP.

Actin in Cells

Intracellular actin cytoskeletal assembly and disassembly are tightly regulated by cell signalling mechanisms. Many signal transduction systems use the actin cytoskeleton as a scaffold, holding them at or near the inner face of the peripheral membrane. This subcellular location allows immediate responsiveness to transmembrane receptor action and the resulting cascade of signal-processing enzymes.

Because actin monomers must be recycled to sustain high rates of actin-based motility during chemotaxis, cell signalling is believed to activate cofilin, the actin-filament depolymerizing protein which binds to ADP-rich actin subunits nearest the filament's pointed-end and promotes filament fragmentation, with concomitant depolymerization in order to liberate actin monomers. In most animal cells, monomeric actin is bound to profilin and thymosin beta-4, both of which preferentially bind with one-to-one stoichiometry to ATP-containing monomers. Although thymosin beta-4 is strictly a monomer-sequestering protein, the behaviour of profilin is far more complex. Profilin enhances the ability of monomers to assemble by stimulating the exchange of actin-bound ADP for solution-phase ATP to yield actin-ATP and ADP. Profilin is transferred to the leading edge by virtue of its PIP_2 binding site, and it employs its poly-L-proline binding site to dock onto end-tracking proteins. Once bound, profilin-actin-ATP is loaded into the monomer-insertion site of actoclampin motors.

Another important component in filament formation is the Arp2/3 complex, which binds to the side of an already existing filament (or "mother filament"), where it nucleates the formation of a new daughter filament at a 70 degree angle relative to the mother filament, effecting a fan-like branched filament network.

Microfilament Associated Proteins

In non-muscle cells, actin filaments are formed proximal to membrane surfaces. Their formation and turnover are regulated by many proteins, including:

- Filament end-tracking protein (e.g., formins, VASP, N-WASP)
- Filament-nucleator known as the Actin-Related Protein-2/3 (or Arp2/3) complex
- Filament cross-linkers (e.g., á-actinin and fascin)
- Actin monomer-binding proteins profilin and thymosin â4
- Filament barbed-end cappers such as Capping Protein and CapG, *etc.*
- Filament-severing proteins like gelsolin.
- Actin depolymerizing proteins such as ADF/cofilin.

The actin filament network in non-muscle cells is highly dynamic. The actin filament network is arranged with the barbed-end of each filament attached to the cell's peripheral membrane by means of clamped-filament elongation motors, the above-mentioned "actoclampins", formed from a filament barbed-end and a clamping protein (formins, VASP, Mena, WASP, and N-WASP). The primary substrate for these elongation motors is profilin-actin-ATP complex which is directly transferred to elongating filament ends. The pointed-end of each filament is oriented toward the cell's interior. In the case of lamellipodial growth, the Arp2/3 complex generates a branched network, and in filopodia a parallel array of filaments is formed.

Actin Acts as a Track for Myosin Motor Motility

Myosin motors are intracellular ATP-dependent enzymes that bind to and move along actin filaments. Various classes of myosin motors have very different behaviours, including exerting tension in the cell and transporting cargo vesicles.

A Proposed Model – Actoclampins Track Filament Ends

One proposed model suggests the existence of actin filament barbed-end-tracking molecular motors termed "actoclampin". The proposed actoclampins generate the propulsive forces needed for actin-based motility of lamellipodia, filopodia, invadipodia, dendritic spines, intracellular vesicles, and motile processes in endocytosis, exocytosis, podosome formation, and phagocytosis. Actoclampin motors also propel such intracellular pathogens as *Listeria monocytogenes*, *Shigella flexneri*, *Vaccinia* and *Rickettsia*. When assembled under suitable conditions, these end-tracking molecular motors can also propel biomimetic particles.

The term actoclampin is derived from *acto-* to indicate the involvement of an actin filament, as in actomyosin, and *clamp* to indicate a clasping device used for strengthening flexible/moving objects and for securely fastening two or more components, followed by the suffix *-in* to indicate its protein origin. An actin filament end-tracking protein may thus be termed a clampin.

Dickinson and Purich recognized that prompt ATP hydrolysis could explain the forces achieved during actin-based motility. They proposed a simple mechanoenzymatic sequence known as the Lock, Load & Fire Model, in which an end-tracking protein remains tightly bound ("locked" or clamped) onto the end of one sub-filament of the double-stranded actin filament. After binding to Glycyl-Prolyl-Prolyl-Prolyl-Prolyl-Prolyl-registers on tracker proteins, Profilin-ATP-actin is delivered ("loaded") to the unclamped end of the other sub-filament, whereupon ATP within the already clamped terminal subunit of the other subfragment is hydrolyzed ("fired"), providing the energy needed to release that arm of the end-tracker, which then can bind another Profilin-ATP-actin to begin a new monomer-addition round.

Steps Involved

The following steps describe one force-generating cycle of an actoclampin molecular motor:

1. The polymerization cofactor profilin and the ATP-actin combine to form a profilin-ATP-actin complex that then binds to the end-tracking unit

2. The cofactor and monomer are transferred to the barbed-end of an actin already clamped filament
3. The tracking unit and cofactor dissociate from the adjacent protofilament, in a step that can be facilitated by ATP hydrolysis energy to modulate the affinity of the cofactor and/or the tracking unit for the filament; and this mechanoenzymatic cycle is then repeated, starting this time on the other sub-filament growth site.

When operating with the benefit of ATP hydrolysis, AC motors generate per-filament forces of 8–9 pN, which is far greater than the per-filament limit of 1–2 pN for motors operating without ATP hydrolysis. The term actoclampin is generic and applies to all actin filament end-tracking molecular motors, irrespective of whether they are driven actively by an ATP-activated mechanism or passively.

Some actoclampins (e.g., those involving Ena/VASP proteins, WASP, and N-WASP) apparently require Arp2/3-mediated filament initiation to form the actin polymerization nucleus that is then "loaded" onto the end-tracker before processive motility can commence. To generate a new filament, Arp2/3 requires a "mother" filament, monomeric ATP-actin, and an activating domain from Listeria ActA or the VCA region of N-WASP. The Arp2/3 complex binds to the side of the mother filament, forming a Y-shaped branch having a 70 degree angle with respect to the longitudinal axis of the mother filament. Then upon activation by ActA or VCA, the Arp complex is believed to undergo a major conformational change, bringing its two actin-related protein subunits near enough to each other to generate a new filament gate. Whether ATP hydrolysis may be required for nucleation and/or Y-branch release is a matter under active investigation.

Peptidoglycan

Peptidoglycan, also known as murein, is a polymer consisting of sugars and amino acids that forms a mesh-like layer outside the plasma membrane of bacteria (but not Archaea), forming the cell wall. The sugar component consists of alternating residues of β-(1,4) linked *N*-acetylglucosamine and *N*-acetylmuramic acid. Attached to the *N*-acetylmuramic acid is a peptide chain of three to five amino acids. The peptide chain can be cross-linked to the peptide chain of another strand forming the 3D mesh-like layer. Some Archaea have a similar layer of pseudopeptidoglycan or pseudomurein, where the sugar residues are β-(1,3) linked *N*-acetylglucosamine and *N*-acetyltalosaminuronic acid. That is why the cell wall of Archaea is insensitive to lysozyme.

Peptidoglycan serves a structural role in the bacterial cell wall, giving structural strength, as well as counteracting the osmotic pressure of the cytoplasm. A common misconception is that peptidoglycan gives the cell its shape; however, whereas peptidoglycan helps maintain the structural strength of the cell, it is actually the MreB protein that facilitates cell shape. Peptidoglycan is also involved in binary fission during bacterial cell reproduction.

The peptidoglycan layer is substantially thicker in Gram-positive bacteria (20 to 80 nanometers) than in Gram-negative bacteria (7 to 8 nanometers), with the attachment of the S-layer. Peptidoglycan forms around 90% of the dry weight of Gram-positive bacteria but only 10% of Gram-negative strains. Thus, presence of high levels of peptidoglycan is the primary determinant of the characterisation of bacteria as gram-positive. In Gram-positive strains, it is important in attachment roles and stereotyping purposes. For both Gram-positive and Gram-negative bacteria, particles of approximately 2 nm can pass through the peptidoglycan.

Structure

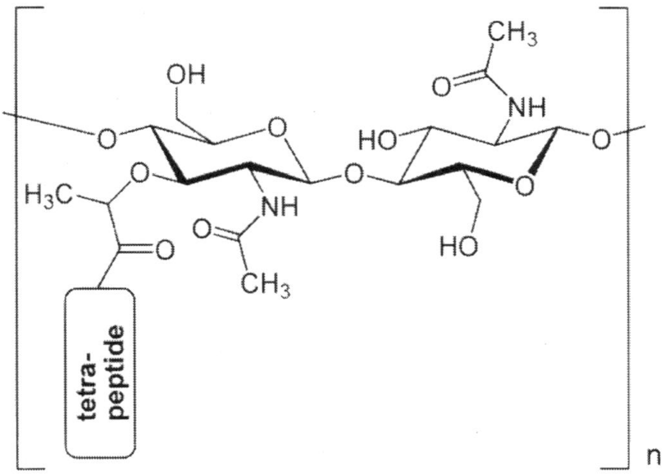

Figure: *Peptidoglycan.*

The peptidoglycan layer in the bacterial cell wall is a crystal lattice structure formed from linear chains of two alternating amino sugars, namely *N*-acetylglucosamine (GlcNAc or NAG) and *N*-acetylmuramic acid (MurNAc or NAM). The alternating sugars are connected by a â-(1,4)-glycosidic bond. Each MurNAc is attached to a short (4- to 5-residue) amino acid chain, containing L-alanine, D-glutamic acid, *meso*-diaminopimelic acid, and D-alanine in the case of *Escherichia coli* (a Gram-negative bacteria) or L-alanine, D-glutamine, L-lysine, and D-alanine with a 5-glycine interbridge between tetrapeptides in the case of *Staphylococcus aureus* (a Gram-positive bacteria). These D-amino acids do not occur in proteins and are thought to help protect against attacks by most peptidases.

Cross-linking between amino acids in different linear amino sugar chains occurs with the help of the enzyme transpeptidase and results in a 3-dimensional structure that is strong and rigid. The specific amino acid sequence and molecular structure vary with the bacterial species.

Antibiotic Inhibition

Some antibacterial drugs such as penicillin interfere with the production of peptidoglycan by binding to bacterial enzymes known as penicillin-binding proteins or transpeptidases. Penicillin-binding proteins form the bonds between oligopeptide crosslinks in peptidoglycan. For a bacterial cell to reproduce through binary fission, more than a million peptidoglycan subunits (NAM-NAG+oligopeptide) must be attached to existing subunits. Mutations in transpeptidases that lead to reduced interactions with an antibiotic are a significant source of emerging antibiotic resistance.

Considered the human body's *own antibiotic,* lysozymes found in tears work by breaking the â-(1,4)-glycosidic bonds in peptidoglycan and thereby destroying many bacterial cells. Antibiotics such as penicillin commonly target bacterial cell wall formation (of which peptidoglycan is an important component) because animal cells do not have cell walls.

Biosynthesis

The peptidoglycan monomers are synthesized in the cytosol and are then attached to a membrane carrier bactoprenol. Bactoprenol transports peptidoglycan monomers across the cell membrane where they are inserted into the existing peptidoglycan.

In the first step of peptidoglycan synthesis, the glutamine, which is an amino acid, donates an amino group to a sugar, fructose 6-phosphate. This turns fructose 6-phosphate into glucosamine-6-phosphate. In step two, an acetyl group is transferred from acetyl CoA to the amino group on the glucosamine-6-phosphate creating *N*-acetyl-glucosamine-6-phosphate. In step three of the synthesis process, the *N*-acetyl-glucosamine-6-phosphate is isomerized, which will change *N*-acetyl-glucosamine-6-phosphate to *N*-acetyl-glucosamine-1-phosphate.

In step 4, the *N*-acetyl-glucosamine-1-phosphate, which is now a monophosphate, attacks UTP. Uridine triphosphate, which is a pyrimidine nucleotide, has the ability to act as an energy source. In this particular reaction, after the monophosphate has attacked the UTP, an inorganic pyrophosphate is given off and is replaced by the monophosphate, creating UDP-N-acetylglucosamine (2,4). (When UDP is used as an energy source, it gives off an inorganic phosphate.) This initial stage, is used to create the precursor for the NAG in peptidoglycan.

In step 5, some of the UDP-N-acetylglucosamine (UDP-GlcNAc) is converted to UDP-MurNAc (UDP-N-acetylmuramic acid) by the addition of a lactyl group to the glucosamine. Also in this reaction, the C3 hydroxyl group will remove a phosphate from the alpha carbon of phosphenol pyruvate. This creates what is called an enol derivative that will be reduced to a "lactyl moiety" by NADPH in step six.

In step 7, the UDP–MurNAc is converted to UDP-MurNAc pentapeptide by the addition of five amino acids, usually including the dipeptide D-alanyl-D-alanine. Each of these reactions requires the energy source ATP. This is all referred to as Stage one.

Stage two occurs in the cytoplasmic membrane. It is in the membrane where a lipid carrier called bactoprenol carries peptidoglycan precursors through the cell membrane. Bactoprenol will attack the UDP-MurNAc penta, creating a PP-MurNac penta, which is now a lipid. UDP-GlcNAc is then transported to MurNAc, creating Lipid-PP-MurNAc penta-GlcNAc, a disaccharide, also a precursor to peptidoglycan. How this molecule is transported through the membrane is still not understood. However, once it is there, it is added to the growing glycan chain. The next reaction is known as tranglycosylation. In the reaction, the hydroxyl group of the GlcNAc will attach to the MurNAc in the glycan, which will displace the lipid-PP from the glycan chain. The enzyme responsible for this is transglycosylase.

Hemicellulose

A hemicellulose (Also known as Polyose) is any of several heteropolymers (matrix polysaccharides), such as arabinoxylans, present along with cellulose in almost all plant cell walls. While cellulose is crystalline, strong, and resistant to hydrolysis, hemicellulose has a random, amorphous structure with little strength. It is easily hydrolyzed by dilute acid or base as well as myriad hemicellulase enzymes.

Composition

Hemicelluloses include xylan, glucuronoxylan, arabinoxylan, glucomannan, and xyloglucan. These polysaccharides contain many different sugar monomers. In contrast, cellulose contains only anhydrous glucose. For instance, besides glucose, sugar monomers in hemicellulose can include xylose, mannose, galactose, rhamnose, and arabinose. Hemicelluloses contain most of the D-pentose sugars, and occasionally small amounts of L-sugars as well. Xylose is in most cases the sugar monomer present in the largest amount, although in softwoods mannose can be the most abundant sugar. Not only regular sugars can be found in hemicellulose, but also their acidified form, for instance glucuronic acid and galacturonic acid can be present.

Structural Comparison to Cellulose

Unlike cellulose, hemicellulose (also a polysaccharide) consists of shorter chains – 500–3,000 sugar units as opposed to 7,000–15,000 glucose molecules per polymer seen in cellulose. In addition, hemicellulose is a branched polymer, while cellulose is unbranched.

Native Structure

Hemicelluloses are embedded in the cell walls of plants, sometimes in chains that form a 'ground' - they bind with pectin to cellulose to form a network of cross-linked fibres.

Biosynthesis

Hemicelluloses are synthesised from sugar nucleotides in the cell's Golgi apparatus. Two models explain their synthesis: 1) a '2 component model' where modification occurs at two transmembrane proteins, and 2) a '1 component model' where modification occurs only at one transmembrane protein. After synthesis, hemicelluloses are transported to the plasma membrane via golgi vesicles.

Applications

As percent content of hemicellulose increases in animal feed, the voluntary feed intake decreases. Hemicellulose is represented by the difference between neutral detergent fibre (NDF) and acid detergent fibre (ADF).

Functions

Microfibrils are cross-linked together by hemicellulose homopolymers. Lignins assist and strengthen the attachment of hemicelluloses to microfibrils.

Hemicellulose from Trees

Hemicellulose found in hardwood trees is predominantly xylan with some glucomannan, while in softwoods it is mainly rich in galactoglucomannan and contains only a small amount of xylan. The average molecular weight is lower than that of cellulose at less than 30,000, as opposed to the 100,000 average molecular weight reported for cellulose.

Pectin

Pectin is a structural heteropolysaccharide contained in the primary cell walls of terrestrial plants. It was first isolated and described in 1825 by Henri Braconnot. It is produced commercially as a white to light brown powder, mainly extracted from citrus fruits, and is used in food as a gelling agent particularly in jams and jellies. It is also used in fillings, medicines, sweets, as a stabilizer in fruit juices and milk drinks, and as a source of dietary fibre.

Biology

In plant biology, pectin consists of a complex set of polysaccharides that are present in most primary cell walls and are particularly abundant in the non-woody parts of terrestrial plants. Pectin is a major component of the middle lamella, where it helps to bind cells together, but is also found in primary cell walls.

The amount, structure and chemical composition of pectin differs among plants, within a plant over time, and in various parts of a plant. Pectin is an important cell wall polysaccharide that allows primary cell wall extension and plant growth. During fruit ripening, pectin is broken down by the enzymes pectinase and pectinesterase, in which process the fruit becomes softer as the middle lamellae break down and cells become separated from each other. A similar process of cell separation caused by the breakdown of pectin occurs in the abscission zone of the petioles of deciduous plants at leaf fall.

Pectin is a natural part of the human diet, but does not contribute significantly to nutrition. The daily intake of pectin from fruits and vegetables can be estimated to be around 5 g (assuming consumption of approximately 500 g fruits and vegetables per day).

In human digestion, pectin binds to cholesterol in the gastrointestinal tract and slows glucose absorption by trapping carbohydrates. Pectin is thus a soluble dietary fibre. Consumption of pectin has been shown to reduce blood cholesterol levels. The mechanism appears to be an increase of viscosity in the intestinal tract, leading to a reduced absorption of cholesterol from bile or food. In the large intestine and colon, microorganisms degrade pectin and liberate short-chain fatty acids that have positive influence on health (prebiotic effect).

Chemistry

Pectins, also known as pectic polysaccharides, are rich in galacturonic acid. Several distinct polysaccharides have been identified and characterised within the pectic group. Homogalacturonans are linear chains of α-(1–4)-linked D-galacturonic acid. Substituted galacturonans are characterized by the presence of saccharide appendant residues (such as D-xylose or D-apiose in the respective cases of xylogalacturonan and apiogalacturonan) branching from a backbone of D-galacturonic acid residues. Rhamnogalacturonan I pectins (RG-I) contain a backbone of the repeating disaccharide: 4)-α-D-galacturonic acid-(1,2)-α-L-rhamnose-(1. From many of the rhamnose residues, sidechains of various neutral sugars branch off. The neutral sugars are mainly D-galactose, L-arabinose and D-xylose, with the types and proportions of neutral sugars varying with the origin of pectin.

Another structural type of pectin is rhamnogalacturonan II (RG-II), which is a less frequent complex, highly branched polysaccharide. Rhamnogalacturonan II is classified by some authors within the group of substituted galacturonans since the rhamnogalacturonan II backbone is made exclusively of D-galacturonic acid units.

Isolated pectin has a molecular weight of typically 60–130,000 g/mol, varying with origin and extraction conditions.

In nature, around 80 percent of carboxyl groups of galacturonic acid are esterified with methanol. This proportion is decreased to a varying degree during pectin extraction. The ratio of esterified to non-esterified galacturonic acid determines the behaviour of pectin in food applications. This is why pectins are classified as high- vs. low-ester pectins (short HM vs. LM-pectins), with more or less than half of all the galacturonic acid esterified.

The non-esterified galacturonic acid units can be either free acids (carboxyl groups) or salts with sodium, potassium, or calcium. The salts of partially esterified pectins are called pectinates, if the degree of esterification is below 5 percent the salts are called pectates, the insoluble acid form, pectic acid.

Some plants such as sugar beet, potatoes and pears contain pectins with acetylated galacturonic acid in addition to methyl esters. Acetylation prevents gel-formation but increases the stabilising and emulsifying effects of pectin.

Amidated pectin is a modified form of pectin. Here, some of the galacturonic acid is converted with ammonia to carboxylic acid amide. These pectins are more tolerant of varying calcium concentrations that occur in use.

To prepare a pectin-gel, the ingredients are heated, dissolving the pectin. Upon cooling below gelling temperature, a gel starts to form. If gel formation is too strong, syneresis or a granular texture are the result, whilst weak gelling leads to excessively soft gels. Pectins gel according to specific parameters, such as sugar, pH and bivalent salts (especially Ca2+).

In high-ester pectins at soluble solids content above 60% and a pH-value between 2.8 and 3.6, hydrogen bonds and hydrophobic interactions bind the individual pectin chains together. These bonds form as water is bound by sugar and forces pectin strands to stick together. These form a 3-dimensional molecular net that creates the macromolecular gel. The gelling-mechanism is called a low-water-activity gel or sugar-acid-pectin gel.

In low-ester pectins, ionic bridges are formed between calcium ions and the ionised carboxyl groups of the galacturonic acid. This is idealised in the so-called "egg box-model". Low-ester pectins need calcium to form a gel, but can do so at lower soluble solids and higher pH-values than high-ester pectins. Normally low-ester pectins form gels with a range of pH from 2.6 to 7.0 and with a soluble solids content between 10 and 70%.

Amidated pectins behave like low-ester pectins but need less calcium and are more tolerant of excess calcium. Also, gels from amidated pectin are thermo-reversible; they can be heated and after cooling solidify again, whereas conventional pectin-gels will afterwards remain liquid.

High-ester pectins set at higher temperatures than low-ester pectins. However, gelling reactions with calcium increase as the degree of esterification falls. Similarly, lower pH-values or higher soluble solids (normally sugars) increase gelling speed. Suitable pectins can therefore be selected for jams and for jellies, or for higher sugar confectionery jellies.

Sources and Production

Apples, guavas, quince, plums, gooseberries, oranges and other citrus fruits, contain large amounts of pectin, while soft fruits like cherries, grapes and strawberries contain small amounts of pectin.

Typical levels of pectin in plants are (fresh weight):

- apples, 1–1.5%
- apricot, 1%
- cherries, 0.4%
- oranges, 0.5–3.5%
- carrots approx. 1.4%
- citrus peels, 30%

The main raw-materials for pectin production are dried citrus peel or apple pomace, both by-products of juice production. Pomace from sugar-beet is also used to a small extent.

From these materials, pectin is extracted by adding hot dilute acid at pH-values from 1.5 – 3.5. During several hours of extraction, the protopectin loses some of its branching and chain-length and goes into solution. After filtering, the extract is concentrated in vacuum and the pectin then precipitated by adding ethanol or isopropanol. An old technique of precipitating pectin with aluminium salts is no longer used (apart from alcohols and polyvalent cations; pectin also precipitates with proteins and detergents).

Alcohol-precipitated pectin is then separated, washed and dried. Treating the initial pectin with dilute acid leads to low-esterified pectins. When this process includes ammonium hydroxide, amidated pectins are obtained. After drying and milling, pectin is usually standardised with sugar and sometimes calcium-salts or organic acids to have optimum performance in a particular application.

Worldwide, approximately 40,000 metric tons of pectin are produced every year.

Uses

The main use for pectin (vegetable agglutinate) is as a gelling agent, thickening agent and stabilizer in food. The classical application is giving the jelly-like consistency to jams or marmalades, which would otherwise be sweet juices. Pectin also reduces syneresis in jams and marmalades and increases the gel strength of low calorie jams. For household use, pectin is an ingredient in gelling sugar (also known as "jam sugar") where it is diluted to the right concentration with sugar and some citric acid to adjust pH. In some countries, pectin is also available as a solution or an extract, or as a blended powder, for home jam making. For conventional jams and marmalades that contain above 60% sugar and soluble fruit solids, high-ester pectins are used. With low-ester pectins and amidated pectins less sugar is needed, so that diet products can be made.

Pectin is used in confectionery jellies to give a good gel structure, a clean bite and it confers a good flavour release. Pectin can also be used to stabilize acidic protein drinks, such as drinking yogurt, to improve the mouth-feel and the pulp stability in juice based drinks and as a fat substitute in baked goods. Typical levels of pectin used as a food additive are between 0.5 and 1.0% – this is about the same amount of pectin as in fresh fruit.

In medicine, pectin increases viscosity and volume of stool so that it is used against constipation and diarrhea. Until 2002, it was one of the main ingredients used in Kaopectate a drug to combat diarrhea, along with kaolinite. Pectin is also used in throat lozenges as a demulcent. In cosmetic products, pectin acts as stabilizer. Pectin is also used in wound healing preparations and speciality medical adhesives, such as colostomy devices.

Yablokov *et al.*, writing in Chernobyl: Consequences of the Catastrophe for People and the Environment, quote research conducted by the Ukrainian Centre of Radiation Medicine and the Belarussian Institute of Radiation Medicine and Endocrinology with the conclusion that "adding pectin preparations to the food of inhabitants of the Chernobyl-contaminated regions promotes an effective excretion of incorporated radionuclides". The authors report on the positive results of using pectin food additive preparations in a number of clinical studies conducted on children in severely polluted areas, with up to 50% improvement over control groups.

In ruminant nutrition, depending on the extent of lignification of the cell wall, pectin is up to 90% digestible by bacterial enzymes. Ruminant nutritionists recommend that the digestibility and energy concentration in forages can be improved by increasing pectin concentration in the forage.

In the cigar industry, pectin is considered an excellent substitute for vegetable glue and many cigar smokers and collectors will use pectin for repairing damaged tobacco wrapper leaves on their cigars.

Pectin is also used in jellybeans.

Legal Status

At the FAO/WHO joint Expert Committee on Food Additives and in the EU, no numerical acceptable daily intake (ADI) has been set, as pectin is considered safe.

In the US, pectin is GRAS – generally recognized as safe. In most foods it can be used according to good manufacturing practices in the levels needed for its application ("quantum satis").

In the International Numbering System (INS), pectin has the number 440. In Europe, pectins are differentiated into the E numbers E440(i) for non-amidated pectins and E440 (ii) for amidated pectins. There are specifications in all national and international legislation defining its quality and regulating its use.

History

Pectin was first isolated and described in 1825 by Henri Braconnot, though the action of pectin to make jams and marmalades was known long before. To obtain well set jams from fruits that had little or only poor quality pectin, pectin-rich fruits or their extracts were mixed into the recipe.

During the industrialization, the makers of fruit preserves soon turned to producers of apple juice to obtain dried apple pomace that was cooked to extract pectin.

Later, in the 1920s and 1930s, factories were built that commercially extracted pectin from dried apple pomace and later citrus-peel in regions that produced apple juice in both the USA and in Europe.

At first, pectin was sold as a liquid extract, but nowadays pectin is often used as dried powder that is easier to store and handle than a liquid.

Organelle

In cell biology, an organelle is a specialized subunit within a cell that has a specific function, and it is usually separately enclosed within its own lipid bilayer.

The name *organelle* comes from the idea that these structures are to cells what an organ is to the body (hence the name *organelle,* the suffix *-elle* being a diminutive). Organelles are identified by microscopy, and can also be purified by cell fractionation. There are many types of organelles, particularly in eukaryotic cells. While prokaryotes do not possess organelles *per se,* some do contain protein-based microcompartments, which are thought to act as primitive organelles.

History and Terminology

In biology *organs* are defined as confined functional units within an organism. The analogy of bodily organs to microscopic cellular substructures is obvious, as from even early works, authors of respective textbooks rarely elaborate on the distinction between the two.

Credited as the first to use a diminutive of *organ* (i.e., little organ) for cellular structures was German zoologist Karl August Möbius (1884), who used the term *organula* (plural of *organulum*, the diminutive of Latin *organum*). From the context, it is clear that he referred to reproduction related structures of protists. In a footnote, which was published as a correction in the next issue of the journal, he justified his suggestion to call organs of unicellular organisms "organella" since they are only differently formed parts of one cell, in contrast to multicellular organs of multicellular organisms. Thus, the original definition was limited to structures of unicellular organisms.

It would take several years before *organulum*, or the later term *organelle*, became accepted and expanded in meaning to include subcellular structures in multicellular organisms. Books around 1900 from Valentin Häcker, Edmund Wilson and Oscar Hertwig still referred to cellular *organs*. Later, both terms came to be used side by side: Bengt Lidforss wrote 1915 (in German) about "Organs or Organells".

Around 1920, the term organelle was used to describe propulsion structures ("motor organelle complex", i.e., flagella and their anchoring) and other protist structures, such as ciliates. Alfred Kühn wrote about centrioles as division organelles, although he stated that, for Vahlkampfias, the alternative 'organelle' or 'product of structural build-up' had not yet been decided, without explaining the difference between the alternatives.

In his 1953 textbook, Max Hartmann used the term for extracellular (pellicula, shells, cell walls) and intracellular skeletons of protists.

Later, the now widely used definition of organelle emerged, after which only cellular structures with surrounding membrane had been considered organelles. However, the more original definition of subcellular functional unit in general still coexists.

In 1978, Albert Frey-Wyssling suggested that the term organelle should refer only to structures that convert energy, such as centrosomes, ribosomes, and nucleoli. This new definition, however, did not win wide recognition.

Types of Organelles

While most cell biologists consider the term organelle to be synonymous with "cell compartment", other cell biologists choose to limit the term organelle to include only those that are DNA-containing, having originated from formerly autonomous microscopic organisms acquired via endosymbiosis.

Under this definition, there would only be two broad classes of organelles (i.e. those that contain their own DNA, and have originated from endosymbiotic bacteria):

- mitochondria (in almost all eukaryotes)
- plastids (e.g. in plants, algae, and some protists).

Other organelles are also suggested to have endosymbiotic origins, but do not contain their own DNA (notably the flagellum).

Under the more restricted definition of membrane-bound structures, some parts of the cell do not qualify as organelles. Nevertheless, the use of organelle to refer to non-membrane bound structures such as ribosomes is common. This has led some texts to delineate between membrane-bound and non-membrane bound organelles. These structures are large assemblies of macromolecules that carry out particular and specialized functions, but they lack membrane boundaries. Such cell structures include:

- large RNA and protein complexes: ribosome, spliceosome, vault
- large protein complexes: proteasome, DNA polymerase III holoenzyme, RNA polymerase II holoenzyme, symmetric viral capsids, complex of GroEL and GroES; membrane protein complexes: photosystem I, ATP synthase
- large DNA and protein complexes: nucleosome
- centriole and microtubule-organizing centre (MTOC)
- cytoskeleton
- flagellum
- cellular structure which is not membrane bound and does not have a well-defined structure: nucleolus

Eukaryotic Organelles

Eukaryotic cells are structurally complex, and by definition are organized, in part, by interior compartments that are themselves enclosed by lipid membranes that resemble the outermost cell membrane. The larger organelles, such as the nucleus and vacuoles, are easily visible with the light microscope. They were among the first biological discoveries made after the invention of the microscope.

Not all eukaryotic cells have each of the organelles listed below. Exceptional organisms have cells that do not include some organelles that might otherwise be considered universal to eukaryotes (such as mitochondria). There are also occasional exceptions to the number of membranes surrounding organelles, listed in the tables below (e.g., some that are listed as double-membrane are sometimes found with single or triple membranes). In addition, the number of individual organelles of each type found in a given cell varies depending upon the function of that cell.

Major eukaryotic organelles				
Organelle	***Main function***	***Structure***	***Organisms***	***Notes***
chloroplast (plastid)	photosynthesis, traps energy from sunlight	double-membrane compartment	plants, protists (rare kleptoplastic organisms)	has some genes; theorized to be engulfed by the ancestral eukaryotic cell (endosymbiosis)
endoplasmic reticulum	translation and folding of new proteins (rough endoplasmic reticulum), expression of lipids (smooth endoplasmic reticulum)	single-membrane compartment	all eukaryotes	rough endoplasmic reticulum is covered with ribosomes, has folds that are flat sacs; smooth endoplasmic reticulum has folds that are tubular
Golgi apparatus	sorting, packaging, processing and modification of proteins	single-membrane compartment	all eukaryotes	cis-face (convex) nearest to rough endoplasmic reticulum; trans-face (concave) farthest from rough endoplasmic reticulum
mitochondria	energy production from the oxidation of glucose substances and the release of adenosine triphosphate	double-membrane compartment	most eukaryotes	has some DNA; theorized to be engulfed by an ancestral eukaryotic cell (endosymbiosis)
vacuole	storage,transportation, helps maintain homeostasis	single-membrane compartment	eukaryotes	
nucleus	DNA maintenance, controls all activities of the cell, RNA transcription	double-membrane compartment	all eukaryotes	contains bulk of genome

Mitochondria and chloroplasts, which have double-membranes and their own DNA, are believed to have originated from incompletely consumed or invading prokaryotic organisms, which were adopted as a part of the invaded cell. This idea is supported in the Endosymbiotic theory.

Minor eukaryotic organelles and cell components			
Organelle/Macromolecule	***Main function***	***Structure***	***Organisms***
acrosome	helps spermatozoa fuse with ovum	single-membrane compartment	many animals
autophagosome	vesicle that sequesters cytoplasmic material and organelles for degradation	double-membrane compartment	all eukaryotes
centriole	anchor for cytoskeleton, organizes cell division by forming spindle fibers	Microtubule protein	animals
cilium	movement in or of external medium; "critical developmental signaling pathway".	Microtubule protein	animals, protists, few plants
eyespot apparatus	detects light, allowing phototaxis to take place		green algae and other unicellular photosynthetic organisms such as euglenids
glycosome	carries out glycolysis	single-membrane compartment	Some protozoa, such as *Trypanosomes*.
glyoxysome	conversion of fat into sugars	single-membrane compartment	plants
hydrogenosome	energy & hydrogen production	double-membrane compartment	a few unicellular eukaryotes
lysosome	breakdown of large molecules (e.g., proteins + polysaccharides)	single-membrane compartment	most eukaryotes
melanosome	pigment storage	single-membrane compartment	animals
mitosome	probably plays a role in Fe-S cluster assembly	double-membrane compartment	a few unicellular eukaryotes that lack mitochondria
myofibril	myocyte contraction	bundled filaments	animals
nucleolus	pre-ribosome production	protein-DNA-RNA	most eukaryotes
parenthesome	not characterized	not characterized	fungi
peroxisome	breakdown of metabolic hydrogen peroxide	single-membrane compartment	all eukaryotes
proteasome	degradation of unneeded or damaged proteins by proteolysis	very large protein complex	All eukaryotes, all archaea, some bacteria
ribosome (80S)	translation of RNA into proteins	RNA-protein	all eukaryotes
vesicle	material transport	single-membrane compartment	all eukaryotes

Other related structures:

- cytosol
- endomembrane system
- nucleosome
- microtubule
- cell membrane

Prokaryotic Organelles

Prokaryotes are not as structurally complex as eukaryotes, and were once thought not to have any internal structures enclosed by lipid membranes. In the past, they were often viewed as having little internal organization; but, slowly, details are emerging about prokaryotic internal structures. An early false turn was the idea developed in the 1970s that bacteria might contain membrane folds termed mesosomes, but these were later shown to be artifacts produced by the chemicals used to prepare the cells for electron microscopy.

However, more recent research has revealed that at least some prokaryotes have microcompartments such as carboxysomes. These subcellular compartments are 100–200 nm in diameter and are enclosed by a shell of proteins. Even more striking is the description of membrane-bound magnetosomes in bacteria, as well as the nucleus-like structures of the *Planctomycetes* that are surrounded by lipid membranes.

Prokaryotic organelles and cell components			
Organelle/Macromolecule	***Main function***	***Structure***	***Organisms***
carboxysome	carbon fixation	protein-shell compartment	some bacteria
chlorosome	photosynthesis	light harvesting complex	green sulfur bacteria
flagellum	movement in external medium	protein filament	some prokaryotes and eukaryotes
magnetosome	magnetic orientation	inorganic crystal, lipid membrane	magnetotactic bacteria
nucleoid	DNA maintenance, transcription to RNA	DNA-protein	prokaryotes
plasmid	DNA exchange	circular DNA	some bacteria
ribosome (70S)	translation of RNA into proteins	RNA-protein	bacteria and archaea
thylakoid	photosynthesis	photosystem proteins and pigments	mostly cyanobacteria
mesosomes	functions of golgi bodies,centrioles etc..	small irregular shaped oraganelle containing ribosomes	present in most of the prokaryotic cells

Proteins and Organelles

The function of a protein is closely correlated with the organelle in which it resides. Some methods were proposed for predicting the organelle in which an uncharacterized protein is located according to its amino acid composition and some methods were based on pseudo amino acid composition.

Turgor Pressure

Turgor pressure pushes the plasma membrane against the cell wall of plant, bacteria, and fungi cells as well as those protist cells which have cell walls. This pressure, turgidity, is caused by the osmotic flow of water from area of low solute concentration outside of the cell into the cell's vacuole, which has a higher solute concentration. Healthy plant cells are turgid and plants rely on turgidity to maintain rigidity. In contrast, this phenomenon is not observed in animal cells which have no cell walls to prevent them from being burst by the flow of water into the cell and must either continually pump out water or live in an isotonic solution where there is no osmotic pressure.

Mechanism

A physical phenomenon known as osmosis causes water to flow from an area of low solute concentration to an area of high solute concentration until the two areas have an equal ratio of solute to water. Normally, the solute diffuses toward equilibrium as well; however, all cells are surrounded by a lipid bilayer cell membrane which permits the flow of water in and out of the cell but restricts the flow of solute under many circumstances. As a result, when a cell is placed in a hypotonic solution, water rushes into the membrane, increasing the cell's volume.

Eventually, the cell's membrane is enlarged such that it pushes against the cell's rigid wall. At this point the cell is said to be turgid. In an isotonic solution, water flows into the cell at the same rate it flows out. The pressure pushing the cell's membrane against its wall is reduced and the cell is said to be flaccid. When a cell is placed in a hypertonic solution water actually flows out of the cell into the surrounding solution. This, plasmolysis, causes the membrane to recede from the wall and is responsible for wilting in plant cells.

Gymnosperm

The gymnosperms are a group of seed-producing plants that includes conifers, cycads, *Ginkgo*, and Gnetales. The term "gymnosperm" comes from the Greek word *gymnospermos*, meaning "naked seeds", after the unenclosed condition of their seeds (called ovules in their unfertilized state). Their naked condition stands in contrast to the seeds and ovules of flowering plants (angiosperms), which are enclosed within an ovary. Gymnosperm seeds develop either on the surface of scales or leaves, often modified to form cones, or at the end of short stalks as in *Ginkgo*.

The gymnosperms and angiosperms together compose the spermatophytes or seed plants. By far the largest group of living gymnosperms is the conifers (pines, cypresses, and relatives), followed by cycads, Gnetophytes (*Gnetum*, *Ephedra* and *Welwitschia*), and *Ginkgo* (a single living species).

Classification

In early classification schemes, the gymnosperms (Gymnospermae) were regarded as a "natural" group. There is conflicting evidence on the question of whether the living gymnosperms form a clade. The fossil record of gymnosperms includes many distinctive taxa that do not belong to the four modern groups, including seed-bearing trees that have a somewhat fern-like vegetative morphology (the so-called "seed ferns" or pteridosperms.) When fossil gymnosperms such as Bennettitales, *Caytonia* and the glossopterids are considered, it is clear that angiosperms are nested within a larger gymnosperm clade, although which group of gymnosperms is their closest relative remains unclear.

Subclass Cycadidae

- Order Cycadales
  - Family Cycadaceae: *Cycas*
  - Family Zamiaceae: *Dioon, Bowenia, Macrozamia, Lepidozamia, Encephalartos, Stangeria, Ceratozamia, Microcycas, Zamia.*

Subclass Ginkgoidae

- Order Ginkgoales
  - Family Ginkgoaceae: *Ginkgo*

Subclass Gnetidae

- Order Welwitschiales
  - Family Welwitschiaceae: *Welwitschia*
- Order Gnetales
  - Family Gnetaceae: *Gnetum*
- Order Ephedrales
  - Family Ephedraceae: *Ephedra*

Subclass Pinidae

- Order Pinales
  - Family Pinaceae: *Cedrus, Pinus, Cathaya, Picea, Pseudotsuga, Larix, Pseudolarix, Tsuga, Nothotsuga, Keteleeria, Abies*
- Order Araucariales
  - Family Araucariaceae: *Araucaria, Wollemia, Agathis*
  - Family Podocarpaceae: *Phyllocladus, Lepidothamnus, Prumnopitys, Sundacarpus, Halocarpus, Parasitaxus,*

Lagarostrobos, Manoao, Saxegothaea, Microcachrys, Pherosphaera, Acmopyle, Dacrycarpus, Dacrydium, Falcatifolium, Retrophyllum, Nageia, Afrocarpus, Podocarpus

- Order Cupressales
  - o Family Sciadopityaceae: *Sciadopitys*
  - o Family Cupressaceae: *Cunninghamia, Taiwania, Athrotaxis, Metasequoia, Sequoia, Sequoiadendron, Cryptomeria, Glyptostrobus, Taxodium, Papuacedrus, Austrocedrus, Libocedrus, Pilgerodendron, Widdringtonia, Diselma, Fitzroya, Callitris* (incl. *Actinostrobus*), *Neocallitropsis, Thujopsis, Thuja, Fokienia, Chamaecyparis, Cupressus, Juniperus, Calocedrus, Tetraclinis, Platycladus, Microbiota*
  - o Family Taxaceae: *Austrotaxus, Pseudotaxus, Taxus, Cephalotaxus, Amentotaxus, Torreya*

Diversity and Origin

There are more than 1000 extant or currently living species of Gymnosperms in 88 plant genera belonging to 14 plant families.

It is widely accepted that the gymnosperms originated in the late Carboniferous period. This appears to have been the result of a whole genome duplication event around 319 million years ago. Early characteristics of seed plants were evident in fossil progymnosperms of the late Devonian period around 380 million years ago. It has been suggested that during the mid-Mesozoic era, pollination of some extinct groups of gymnosperms were by extinct species of scorpionflies that had specialized proboscis for feeding on pollination drops. The scorpionflies likely engaged in pollination mutualisms with gymnosperms, long before the similar and independent coevolution of nectar-feeding insects on angiosperms.

Conifers are by far the most abundant extant group of gymnosperms with six to eight families, with a total of 65-70 genera and 600-630 species (696 accepted names). Conifers are woody plants and most are evergreens. The leaves of many conifers are long, thin and needle-like, others species, including most Cupressaceae and some Podocarpaceae, have flat, triangular scale-like leaves. *Agathis* in Araucariaceae and *Nageia* in Podocarpaceae have broad, flat strap-shaped leaves. Cycads are the next most abundant group of gymnosperms, with two or three families, 11 genera, and approximately 300 species. The other extant groups are the 75-80 species of Gnetales and one species of Ginkgo.

Bibliography

Alberts B, Johnson A, Lewis J: *Molecular Biology of the Cell,* Garland Science, New York, 2002.

Alberts, Bruce: *Molecular Biology of the Cell,* Garland, New York, 2002.

Alberts, Bruce; Johnson, Alexander; Lewis, Julian; Raff, Martin; Roberts, Keith; Walter, Peter: *Molecular Biology of the Cell,* Garland Science, New York, 2002.

Alexopoulos, C. J., C. W. Mims, & M. Blackwell: *Introductory Mycology,* John Wiley & Sons, New York, 1996.

Bhattacharya, D.: *Origins of Algae and their Plastids*, Springer-Verlag/Wein, New York, 1997.

Brock, Thomas D., Michael T. Madigan, John M. Martinko, & Jack Parker: *Biology of Microorganisms*, Prentice Hall, Englewood Cliffs, NJ, 1994.

Bruce Alberts: *Molecular Biology of the Cell,* Garland Science, New York, 2002.

Burgess,, Jeremy: *An Introduction to Plant Cell Development,* Cambridge University Press, Cambridge, 1989.

Curtis, M.: *The Geometry of DNA: A Structural Revision,* Blue Gallery, London, 1996.

Doijode S. D.: *Seed Germination in Fruits*, New Delhi, Malhotra Publishers, 1993.

Fortuner, R.: *Nematode Identification and Expert System Technology,* New York, Plenum Press, 1988.

Howland, John L.: *The Surprising Archaea: Discovering Another Domain of Life*, Oxford University Press, Oxford, 2000.

Hudler, George W.: *Magical Mushrooms, Mischievous Molds*, Princeton University Press, Princeton, NJ, 1998.

John, D.M.; Brook, A.J.; Whitton, B.A.: *The Freshwater Algal Flora of the British Isles: An Identification Guide to Freshwater and Terrestrial Algae,* Cambridge University Press, Cambridge 2002.

Kolattukudy, PE: *Plant Cuticles*, BIOS Scientific Publishers Ltd., Oxford, 1996.

Leigh, RA and D Sanders: *The Plant Vacuole*, Academic Press, California and London, 1997.

Lodish H, Berk A, Zipursky LS: *Molecular Cell Biology,* Scientific American Books, New York, 2004.

Margulis, L.: *Origin of Eukaryotic Cells*, Yale University Press, New Haven, 1970.

Nobel, P. S.: *Physicochemical and Environmental Plant Physiology*, Academic Press, San Diego, 1999.

Oparka, K. J.: *Plasmodesmata,* Blackwell Pub Professional, NJ, 2005.

Raper, Kenneth B.: *The Dictyostelids*, Princeton University Press, Princeton, NJ, 1984.

Raven, PH; Evert RF, Eichhorm SE: *Biology of Plants*, WH Freeman, New York, 1999.

Ruse, Michael, and Christopher A. Pynes: *The Stem Cell Controversy: Debating the Issues*, Amherst, NY: Prometheus, 2003.

Sanchez, P. A.: *Properties and Management of Soils in the Tropics*, John Wiley and Sons, New York, 1976.

Singleton P: *Bacteria in Biology, Biotechnology and Medicine,* Wiley, New York, 1999.

Steer, Brian E.S. Gunning, Martin W.: *Plant Cell Biology: Structure and Function*, Jones and Bartlett Publishers, Boston, Mass, 1996.

Tyree, MT; MH Zimmermann: *Xylem Structure and the Ascent of Sap*, Springer-Verlag, New York USA, 2003.

Venes, Donald: *Taber's Cyclopedic Medical Dictionary*, F.A. Davis Company, Philadelphia, 2001.

William H. Elliott and Daphne C. *: Biochemistry and Molecular Biology*, Oxford University Press, Delhi, 2005.

Wise, R.R.; Hoober, J.K.: *The Structure and Function of Plastids*, Springer, NY, 2007.

Index

❑❑❑